Transboundary Floods: Reducing Risks through Flood Management

NATO Science Series

A Series presenting the results of scientific meetings supported under the NATO Science Programme.

The Series is published by IOS Press, Amsterdam, and Springer in conjunction with the NATO Public Diplomacy Division

Sub-Series

I. Life and Behavioural Sciences	IOS Press
II. Mathematics, Physics and Chemistry	Springer
III. Computer and Systems Science	IOS Press
IV. Earth and Environmental Sciences	Springer

The NATO Science Series continues the series of books published formerly as the NATO ASI Series.

The NATO Science Programme offers support for collaboration in civil science between scientists of countries of the Euro-Atlantic Partnership Council. The types of scientific meeting generally supported are "Advanced Study Institutes" and "Advanced Research Workshops", and the NATO Science Series collects together the results of these meetings. The meetings are co-organized bij scientists from NATO countries and scientists from NATO's Partner countries – countries of the CIS and Central and Eastern Europe.

Advanced Study Institutes are high-level tutorial courses offering in-depth study of latest advances in a field.

Advanced Research Workshops are expert meetings aimed at critical assessment of a field, and identification of directions for future action.

As a consequence of the restructuring of the NATO Science Programme in 1999, the NATO Science Series was re-organised to the four sub-series noted above. Please consult the following web sites for information on previous volumes published in the Series.

http://www.nato.int/science
http://www.springer.com
http://www.iospress.nl

Series IV: Earth and Environmental Sciences – Vol. 72

Transboundary Floods: Reducing Risks Through Flood Management

edited by

Jiri Marsalek
National Water Research Institute, Burlington,
Ontario, Canada

Gheorghe Stancalie
National Meteorological Administration,
Bucharest, Romania

and

Gabor Balint
Environmental and Water Management Research Institute,
Budapest, Hungary

 Springer

Published in cooperation with NATO Public Diplomacy Division

Proceedings of the NATO Advanced Research Workshop on
Transboundary Floods: Reducing Risks and Enhancing Security
Through Improved Flood Management Planning
Baile Felix (Oradea), Romania
4–8 May 2005

A C.I.P. Catalogue record for this book is available from the Library of Congress.

ISBN-10 1-4020-4901-3 (PB)
ISBN-13 978-1-4020-4901-9 (PB)
ISBN-10 1-4020-4900-5 (HB)
ISBN-13 978-1-4020-4900-2 (HB)
ISBN-10 1-4020-4902-1 (e-book)
ISBN-13 978-1-4020-4902-6 (e-book)

Published by Springer,
P.O. Box 17, 3300 AA Dordrecht, The Netherlands.

www.springer.com

Printed on acid-free paper

Library
University of Texas
at San Antonio

All Rights Reserved
© 2006 Springer
No part of this work may be reproduced, stored in a retrieval system, or transmitted in any form or by any means, electronic, mechanical, photocopying, microfilming, recording or otherwise, without written permission from the Publisher, with the exception of any material supplied specifically for the purpose of being entered and executed on a computer system, for exclusive use by the purchaser of the work.

Printed in the Netherlands.

TABLE OF CONTENTS

Preface	ix
Acknowledgement	xi
List of Participants	xiii

CHAPTER 1 COLLECTION AND TRANSMISSION OF DATA USED IN FLOOD MANAGEMENT

MODIS-BASED FLOOD DETECTION, MAPPING AND MEASUREMENT: THE POTENTIAL FOR OPERATIONAL HYDROLOGICAL APPLICATIONS 1
R. Brakenridge and E. Anderson

EXPERIENCE WITH DISCHARGE MEASUREMENTS DURING EXTREME FLOOD EVENTS 13
J. Szekeres

DEVELOPMENT OF THE HYDROMETEOROLOGICAL TELEMETRY SYSTEM IN THE KÖRÖS RIVER BASIN 23
A. Kiss and B. Lukács

EXPERIENCE FROM OPERATION OF THE JOINT HUNGARIAN-UKRAINIAN HYDROLOGICAL TELEMETRY SYSTEM OF THE UPPER TISZA 33
K. Konecsny

LAND USE MAP FROM ASTER IMAGES AND WATER MASK ON MODIS IMAGES 45
J. Kerényi and M. Putsay

CHAPTER 2 FLOOD FORECASTING AND MODELLING

APPLICATION OF METEOROLOGICAL ENSEMBLES FOR DANUBE FLOOD FORECASTING AND WARNING 57
G. Bálint, A. Csík, P. Bartha, B. Gauzer and I. Bonta

COUPLING THE HYDROLOGIC MODEL CONSUL AND THE
METEOROLOGICAL MODEL HRM IN THE CRISUL ALB AND
CRISUL NEGRU RIVER BASINS 69
R. Mic, C. Corbus, V.I. Pescaru and L. Velea

ROUTING OF NUMERICAL WEATHER PREDICTIONS
THROUGH A RAINFALL-RUNOFF MODEL 79
K. Hlavcova, J. Szolgay, R. Kubes, S. Kohnova and M. Zvolenský

THEORETICAL GROUND OF NORMATIVE BASE FOR
CALCULATION OF THE CHARACTERISTICS OF THE
MAXIMUM RUNOFF AND ITS PRACTICAL REALISATION 91
E. Gopchenko and V. Ovcharuk

SCENARIOS OF FLOOD REGIME CHANGES DUE TO LAND
USE CHANGE IN THE HRON RIVER BASIN 99
Z. Papankova, O. Horvat, K. Hlavcova, J. Szolgay and S. Kohnova

BUNDLED SOFTWARE FOR LONG-TERM TERRITORIAL
FORECASTS OF SPRING FLOODS 111
E. Gopchenko and J. Shakirzanova

A STOCHASTIC APPROACH TO FLOOD WAVE
PROPAGATION ON THE CRISUL ALB RIVER 121
R. Drobot and C. Ilinca

SIMULATION OF FLOODING DUE TO THE CRISUL ALB
DYKE FAILURE DURING THE APRIL 2000 FLOOD 133
A. Nitu, R. Mic and R. Amaftiesei

MATHEMATICAL MODELLING OF FLASH FLOODS IN
NATURAL AND URBAN AREAS 143
M. Szydlowski

FLOOD MODELLING CONCEPT AND REALITY - AUGUST
2002 FLOOD IN THE CZECH REPUBLIC 155
P. Sklenář, E. Zeman, J. Špatka and P. Tachecí

SIMULATION OF THE SUPERIMPOSITION OF FLOODS IN
THE UPPER TISZA REGION 171
J. Szilágyi, G. Bálint, A. Csík, B. Gauzer, and M. Horoszné-Gulyás

HARMONISING QUALITY ASSURANCE IN MODEL-
BASED STUDIES OF CATCHMENT AND RIVER BASIN
MANAGEMENT   183
J. Spatka

CHAPTER 3   FLOOD MANAGEMENT

RESEARCH, EDUCATION AND INFORMATION SYSTEMS
IN THE CONTEXT OF A FRAMEWORK FOR FLOOD
MANAGEMENT   193
R.K. Price

OVERVIEW OF THE NATO SCIENCE FOR PEACE PROJECT
ON MANAGEMENT OF TRANSBOUNDARY FLOODS IN THE
CRISUL-KÖRÖS RIVER SYSTEM   205
J. Marsalek, G. Stancalie, R. Brakenridge, M. Putsay, R. Mic and
J. Szekeres

COPING WITH UNCERTAINTIES IN FLOOD MANAGEMENT   219
I. Bogardi

TRANSBOUNDARY FLOODS IN AZERBAIJAN   231
R. Mammedov

FLOOD DEFENCE BY MEANS OF COMPLEX STRUCTURAL
MEASURES   237
S. Pagliara

DYKE FAILURES IN HUNGARY OF THE PAST 220 YEARS   247
S. Tóth and L. Nagy

EFFECT OF THE SALARD TEMPORARY STORAGE
RESERVOIR ON THE BARCAU RIVER FLOWS NEAR
THE ROMANIA-HUNGARY BORDER   259
A. Purdel and P. Mazilu

FLOOD CONTROL MANAGAMENT WITH SPECIAL
REFERENCE TO EMERGENCY RESERVOIRS   265
Z. Galbáts

STRUCTURAL FLOOD CONTROL MEASURES IN THE
CRISUL REPEDE BASIN AND THEIR EFFECTS IN ROMANIA
AND HUNGARY   277
M. Tentis, M. Gale and C. Morar

CONTRIBUTION OF EARTH OBSERVATION DATA
SUPPLIED BY THE NEW SATELLITE SENSORS TO
FLOOD MANAGEMENT 287
G. Stancalie, S. Catana, A. Irimescu, E. Savin, A. Diamandi,
A. Hofnar and S. Oancea

ON-LINE SUPPORT SYSTEM FOR TRANSBOUNDARY
FLOOD MANAGEMENT: DESIGN AND FUNCTIONALITY 305
V. Craciunescu and G. Stancalie

TERRITORIAL FLOOD DEFENCE: A ROMANIAN
PERSPECTIVE 315
M. Lucaciu

INDEX 335

PREFACE

Flood damages are increasing worldwide as a result of frequent recurrence of large floods in many parts of the world, existing and continuing encroachment on flood plains and aging flood protection structures. In the aftermath of recent flood events, the public and experts are looking for ways of protecting life, land, property and the environment, and reducing flood damages. Towards this end, many flood management measures have been practiced, including living with floods, non-structural measures (e.g., regulations, flood defence by flood forecasting and warning, evacuations, and flood insurance), and structural measures (e.g., land drainage modifications, reservoirs, dykes and polders). Such flood management is difficult in river basins controlled by a single authority, and becomes even more challenging when dealing with transboundary floods, which may originate in one country and then propagate downstream to another country, or countries. Under such circumstances, the demands on information and data sharing, and close collaboration in all aspects of flood management are particularly strong and important. Recognising the challenge of transboundary floods, the workshop organisers proposed to the NATO Collaborative Science Program to hold a workshop on this topic. After receiving the NATO grant, the NATO Advanced Research Workshop on Transboundary Floods: Reducing Risks and Enhancing Security was held in Oradea, Romania, from May 4 to 8, 2005.

Preparatory activities included recruiting keynote speakers, selecting workshop participants, finalising the workshop programme, and holding the workshop in Hotel Termal at Baile Felix (Oradea). There were 49 full-time participants from 17 countries at the workshop, and additional observers also audited the workshop program. Extensive experience of workshop participants in this field is reflected in the workshop proceeding, which comprise almost 30 selected papers. Finally, whenever trade, product or firm names are used in the proceedings, it is for identification and descriptive purposes only, without implying endorsement by the Editors, Authors or NATO.

The proceedings that follow reflect only the formal workshop presentations. Besides these presentations, posters, and extensive formal discussions, there were many other ways of sharing and exchanging information among the participants, in the form of new or renewed collaborative links, professional networking and personal friendships. The peaceful atmosphere of the spa resort Hotel Termal contributed to the success of this workshop. For this success, the editors and organisers are indebted to many who helped stage the workshop and produce its proceedings, as listed in the Acknowledgement.

Jiri Marsalek
Burlington, Ontario, Canada

Gheorghe Stancalie
Bucharest, Romania

Gabor Balint
Budapest, Hungary

ACKNOWLEDGEMENT

This Advanced Research Workshop (ARW) resulted from hard work of many individuals and organisations. The workshop was proposed and directed by Dr. Jiri Marsalek, National Water Research Institute (NWRI), Environment Canada, Burlington, Canada, and Dr. Gheorghe Stancalie, Romanian Meteorological Administration, Bucharest, Romania. They were assisted by three other members of the workshop Organising Committee: Gabor Balint, National Forecasting Services, Environmental and Water Management Research Institute, Budapest, Hungary; Mitrut Tentis, Crisuri River Authority, Oradea, Romania; and, Evzen Zeman, DHI Hydroinform a.s., Prague, Czech Republic.

The ARW was sponsored by NATO, Public Diplomacy Division, Collaborative Programmes Section, in the form of a grant; by the employers of the members of the Organising Committee, who provided additional resources required to prepare and run the workshop and preprint and publish its proceedings.

The workshop preparatory work and secretariat services were provided by Corina Alecu and Anisoara Irimescu, Romanian Meteorological Administration, Bucharest, Romania. Local workshop arrangements in Baile Felix were done by a team from the Crisuri Water Authority, led by Octavian Streng, who was assisted by M. Gale, C. Morar and others.

The editing of proceedings was done by Jiri Marsalek, Gheorghe Stancalie and Gabor Balint, and the camera ready manuscript was prepared by Quintin Rochfort, National Water Research Institute, Burlington, Canada.

Special thanks are due to Dr. D. Beten, Programme Director, Environmental Security, NATO, who provided liaison between the workshop organisers and NATO, and personally assisted with many tasks.

Finally, the organisers are indebted to all the above contributors and, above all, to the participants, who made this workshop a memorable interactive learning experience for all.

LIST OF PARTICIPANTS

Directors

Marsalek, J. National Water Research Institute
867 Lakeshore Road, Burlington, ON L7R 4A6
CANADA

Stancalie, G. Romanian Meteorological Administration
97 Sos. Bucuresti-Ploiesti, sector 1, 013686, Bucharest
ROMANIA

Key Speakers

Verdiyev, R. ECORES, NGO
36 Huseynbala Aliyev Str. Apt. 52, Microdistrict,
Baku
AZERBAIJAN

Zeman, E. DHI Hydroinform a.s.
Na Vrsich 5, Prague 100 00
CZECH REPUBLIC

Spatka, J. DHI Hydroinform a.s.
Na Vrsich 5, Prague 100 00
CZECH REPUBLIC

Balint, G. VITUKI,
Kvassay J. u. 1. H-1095 Budapest
HUNGARY

Price, R.K. UNESCO-IHE
Westvest 7, PO Box 3015, Delft 2601 DA
NETHERLANDS

Drobot, R. Technical Univ. of Civil Eng. Bucharest
B-dul Lacul Tei, NR.124, sector 2, RO-72302
Bucharest
ROMANIA

Lucaciu, M.	Arad County Emergency Situation Inspectorate Eftimie Murgu Street 3-5, 2900 Arad ROMANIA
Saul, A.J.	University of Sheffield Mappin Street, Sheffield, S1 3JD UK
Bogardi, I.	University of Nebraska W359 Nebraska Hall, Lincoln NE 68588-05 USA
Mavlyanov, N.	Institute Hydroengeo, State Committee on Geology and Mineral Resources 64, N. Hodjibaev Str., Mirzo Ulugbek district Taskent 700041 UZBEKISTAN

Other Participants

Hakopian, C.	Department of Physical Geography, Yerevan State University 1 Alek Manoukian Street, 375025 Yerevan REPUBLIC OF ARMENIA
Vardanian, T.	Department of Physical Geography, Yerevan State University 1 Alek Manoukian Street, 375025 Yerevan REPUBLIC OF ARMENIA
Mustafayev, I.	Institute of Radiation Problems, Azerbaijan National Academy of Sciences 31a H.Javid ave, Baku, AZ1143 AZERBAIJAN REPUBLIC
Mammedov, R.	Department of Environment, Association of International Hydrological Programme 4 Academician Hasan Aliyev Street, 190, Baku, AZ1065 AZERBAIJAN REPUBLIC

LIST OF PARTICIPANTS

Kukharchyk, T.	Institute for Problems of Natural Resources Use & Ecology of National Academy of Belarus Staroborysovski tract, 10, Minsk, 220114 BELARUS
Otsla, J.	Parnu County Rescue Service Pikk 20, Parnu 80010 ESTONIA
Luik, T.	Parnu County Rescue Service Pikk 20, Parnu 80010 ESTONIA
Csík, A.	VITUKI Kvassay J. u. 1. H-1095 Budapest HUNGARY
Kiss, A.	Koros Region Environment and Water Directorate Városház u. 26, H-5700 Gyula HUNGARY
Szekeres, J.	VITUKI, Kvassay J. u. 1. H-1095 Budapest HUNGARY
Galbats, Z.	Koros Region Environment and Water Directorate Városház u. 26, H-5700 Gyula HUNGARY
Konecsny, K.	VITUKI, Kvassay J. u. 1. H-1095 Budapest HUNGARY
Luidort, A.	Upper Tisza Valley Environment and Water Directorate 19 Szecheny, 4700 Nyiregyhaza HUNGARY
Horosz-Gulyás, M.	VITUKI, Kvassay J. u. 1. H-1095 Budapest HUNGARY

LIST OF PARTICIPANTS

Putsay, M.	Hungarian Meteorological Service Kitaibel Pal Street 1, 1024 Budapest HUNGARY
Toth, S.	National Directorate for Environment, Nature and Water Marvany u. 1/c., Budapest H 1012 HUNGARY
Pagliara, S.	University of Pisa Via Gabba 22, Pisa 56100 ITALY
Szydlowski, M.	Gdansk University of Technology Narutowicza 11/12, Gdansk 80-952 POLAND
Alecu, C.	National Meteorological Administration 97 Sos. Bucuresti-Ploiesti, sector 1, 013686, Bucharest ROMANIA
Irimescu, A.	National Meteorological Administration 97 Sos. Bucuresti-Ploiesti, sector 1, 013686, Bucharest ROMANIA
Craciunescu, V.	National Meteorological Administration 97 Sos. Bucuresti-Ploiesti, sector 1, 013686, Bucharest ROMANIA
Mic, R.	National Institute of Hydrology 97 Sos. Bucuresti-Ploiesti, sector 1, 013686, Bucharest ROMANIA
Tentis, M.	Crisuri Rivers Authority Ion Bogdan Street 35, 3700 Oradea ROMANIA

LIST OF PARTICIPANTS

Nitu, A.	National Institute of Hydrology 97 Sos. Bucuresti-Ploiesti, sector 1, 013686, Bucharest ROMANIA
Pescaru, V.	National Meteorological Administration 97 Sos. Bucuresti-Ploiesti, sector 1, 013686, Bucharest ROMANIA
Purdel, A.	National Institute of Hydrology 97 Sos. Bucuresti-Ploiesti, sector 1, 013686, Bucharest ROMANIA
Streng, O.	Crisuri Rivers Authority Ion Bogdan Street 35, 3700 Oradea ROMANIA
Hlavcova, K.	Slovak Technical University Radlinského 11, Bratislava 813 68 SLOVAKIA
Horvat, O.	Slovak Technical University Radlinského 11, Bratislava 813 68 SLOVAKIA
Papankova, Z.	Slovak Technical University Radlinského 11, Bratislava 813 68 SLOVAKIA
Anderson, E.	Dartmouth Flood Observatory, Dartmouth College Hinman Box 6017, Hanover, NH 03755 USA
Shakirzanova, J.	Odessa State Environmental University 15 Lvovskaya Street, Odessa 65016 UKRAINE
Ivashchuk, O.	Remote Sensing Laboratory, Ukrainian Hydrometeorological Research Institute 37 Nauki str., Kyiv 03028 UKRAINE

LIST OF PARTICIPANTS

Kryvobok, O.	Remote Sensing Laboratory, Ukrainian Hydrometeorological Research Institute 37 Nauki str., Kyiv 03028 UKRAINE
Ovcharuk, V.	Odessa State Environmental University 15 Lvovskaya street, Odessa 65016 UKRAINE
Frank, L.	State Committee for Nature Protection A. Timur 99, Tashkent 700084 UZBEKISTAN

CHAPTER 1 COLLECTION AND TRANSMISSION
OF DATA USED IN FLOOD MANAGEMENT

MODIS-BASED FLOOD DETECTION, MAPPING AND MEASUREMENT: THE POTENTIAL FOR OPERATIONAL HYDROLOGICAL APPLICATIONS

R. BRAKENRIDGE[1]
Dartmouth Flood Observatory, Department of Geography, Dartmouth College, Hanover, NH, 03755, USA

E. ANDERSON
Dartmouth Flood Observatory, Department of Geography, Dartmouth College, Hanover, NH, 03755, USA

Abstract. The internationally available and free rapid response data from NASA's two MODIS sensors have considerable potential for operational applications in applied hydrology, including 1) flood detection, characterisation, and warning, 2) flood disaster response and damage assessment, and 3) flood disaster prevention or mitigation. Each requires different strategies for operational implementation. Successful transition to routine use requires start-up investments in personnel time, training, computing, and other infrastructure. However, because comparable follow-on sensors are already planned (e.g. the NASA/NOAA VIIRS sensor aboard NPOESS), such investment can provide permanent enhancements to hydrological measurement and forecast capabilities. In regions where rivers and streams cross international boundaries, MODIS flood detection capabilities are especially useful: they provide consistent and independently verifiable information. However, even within relatively well-gauged regions, such as the U.S., the capability to characterise inundation as it occurs is an important and economical enhancement to flood warning and flood response. We provide pilot study examples for a region entirely

---

[1] To whom all correspondence should be addressed. G. Robert Brakenridge, Dartmouth Flood Observatory, Dept. of Geography, Dartmouth College, Hanover, NH, 03755, USA; e-mail: G.Robert.Brakenridge@Dartmouth.EDU

within the central U.S and also an international transboundary region within Eastern Europe.

Keywords: MODIS, remote sensing, floods, rivers, disaster response, flood hazard

1. Introduction

The two MODIS (Moderate Resolution Imaging Spectroradiometer) sensors aboard the U.S. satellites Terra and Aqua are nearly ideal region-scale flood mapping and surface water measurement tools (Brakenridge et al. 2003). Operating since early 2001, MODIS data include numerous spectral bands at 500 m and 1 km (at nadir) spatial resolution, but also two (visible and near IR) spectral bands at 250 m resolution. These latter provide excellent water/land discrimination in many settings at acceptable spatial resolution for many applications.

The georeferencing data provided with each "Level 1b swath" scene are accurate to +/- 50 m: this means that the satellite data can be transformed, in "batch mode" (without further human attention) to georectified image-maps, in user-specified map projection. Subsequent image maps prepared for the same areas are, in our experience, in nearly exact registration, so that image change detection approaches can be economically employed and provide maximum sensitivity to small surface water changes.

MODIS data offer frequent (more than daily) coverage, and are provided by NASA free to the international public via the world-wide-web at the following two locations: http://delenn.gsfc.nasa.gov/~imswww/pub/-imswelcome/, and http://rapidfire.sci.gsfc.nasa.gov/. Typical individual scene swath file sizes are 250 megabytes; swath width is 2330 km; only the central portions of the swath provide the full spatial resolution. Recent versions of Envi™, among other commercial remote sensing software, can read, rectify, and re-project MODIS data without further modification. Also, unsupervised classification algorithms supported by Envi™ can identify water image pixels consistently, and groups of water pixels can be translated via vectorisation algorithms into GIS vectors, or outlines. This allows MODIS-observed water to be exported as map layers and integrated with a wide variety of other map displays. The success in water classification is largely due to spectral characteristics of the band 2 data (841 - 876 nm); this band was in fact chosen by the sensor design team to provide excellent water/land discrimination.

Because the data are freely available, and all data obtained since launch are also archived and accessible, and because MODIS data are frequent, well-calibrated, and of spatial resolution adequate to map many small and large river floods, there is high potential for operational applications in applied hydrology (Marsalek et al. 2004). These include: 1) flood detection, characterisation, and warning, 2) flood disaster response and damage assessment, and 3) flood disaster prevention or mitigation. Each requires different strategies for operational implementation.

2. Flood detection, measurement, and warning

2.1. GENERAL COMMENTS

The first area of potential operational use of MODIS also requires the most investment in developing appropriate processing methodologies, including analysis and display techniques. The practical challenge is to accomplish routine and frequent comparison of new MODIS 250 m image data to previous scenes: in order to discover newly flooded areas and to measure the changes.

There is no unique spectral signature for floods (Mertes et al. 2004). However, topography and the record of prior events provide important spatial restrictions to the areas of interest for future flooding. It is possible, using this record, to establish "alarm" thresholds at these locations, wherein particular changes in the remote sensing signal reliably indicate the onset of flooding (Brakenridge et al. 2005).

On a pilot basis, we have begun testing the following work steps towards this purpose. The labour needed depends very much on the total land area to be included for frequent observation. We have implemented this approach for two regions (Figs. 1, 2) measuring approximately 300 x 300 km: in the central U.S. (central Indiana and Illinois), and in central Europe (eastern Hungary, southern Ukraine, and western Romania).

The following tasks must be accomplished: 1) preparation of the initial "wide area hydrological monitoring display", 2) collection of MODIS time series data, and 3) initiation of operational monitoring over the defined area and for flood detection purposes.

2.2. PREPARATION OF WIDE AREA HYDROLOGICAL MONITORING DISPLAYS

The tasks include:

- An initial reference MODIS 250 m image covering the region is obtained, rectified, and geocoded (Figs. 1, 2).
- The complete time series of MODIS data is selected (cloudy scenes are not used) and then obtained via ftp from web site data distribution

locations. These are each geocoded and co-registered to the initial scene.

- Water pixel "classification" is accomplished using a threshold approach and "NDVI" band ratio (band 2-band 1/band 2+band 1) values, for typical or average conditions, and also for flooded conditions. The latter will commonly require several scenes obtained over a period of several days to provide complete coverage, due to cloud cover.
- Vectorisation of high water limits and low water limits is then accomplished. The resulting GIS vector files are then incorporated into the wide area monitoring display.

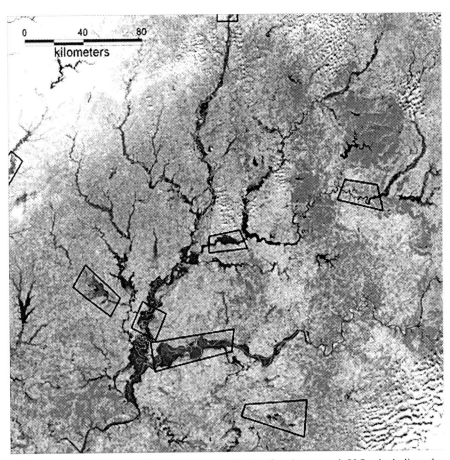

Fig. 1. Wide-area hydrological monitoring display for the central U.S., including the Wabash, Little Wabash, and White Rivers. Gauging reach locations are shown as black outlines. The region was experiencing significant flooding when this MODIS band 2 scene was acquired: January 14, 2005

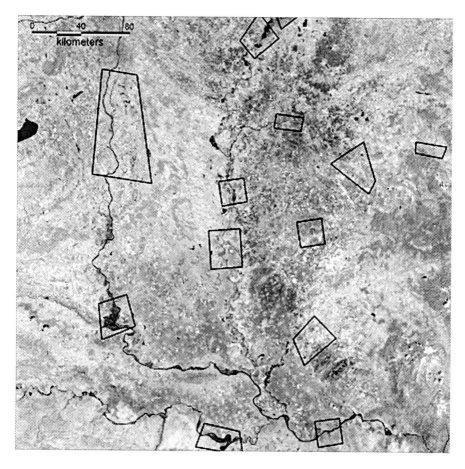

Fig. 2. Wide-area hydrological monitoring display for a portion of Eastern Europe, including the Danube and Tisza river valleys. Gauging reach locations are shown as black outlines. The region was experiencing significant flooding when this MODIS band 2 scene was acquired: April 3, 2005

2.3. COLLECTION OF MODIS TIME SERIES DATA

The tasks include:

- Gauging reach locations are selected and defined. Measurement subreaches (areas sometimes flooded within each reach) and calibration subreaches (areas within the reach but not subject to flooding) are also defined for each reach (Fig. 3).

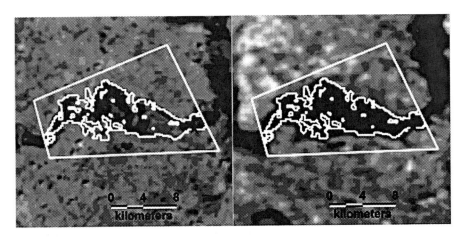

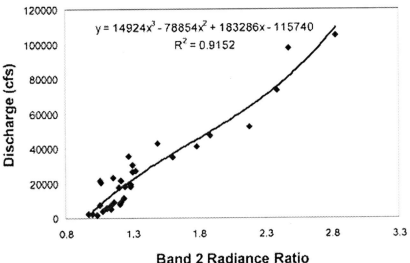

Fig. 3. Top: MODIS band 2 images during minor flooding (left, July 17, 2003, 42,900 cfs) and major flooding (right, January 13, 2004, 97,600 cfs), visually showing surface water changes along a White River, Indiana, USA gauging reach. Bottom: MODIS band 2 calibrated radiance ratios from this reach versus the discharge measured at a US Geological Survey gauging station at Petersburg (located at east end of gauging reach)

- Either band 2 average reflectance or band 2 average radiance values are retrieved for the assembled time series, and for both the measurement and the calibration subreaches. A ratio time series is calculated for each reach. Lower radiance or reflectance values in band 2 are associated with larger water surface area. Therefore, higher values for the ratio (calibration reach/measurement reach) indicate higher reach water surface areas, higher river stages, and, thus, higher river discharges (Brakenridge et al. 2005; Brakenridge et al. 2003).

- Ratio values associated with "bankfull" conditions, immediately below overbank flooding, are estimated by comparison to the associated classified water pixels.
- The time series of overbank floods at each gauging reach can now be determined by the dates of each exceedance of this alarm threshold.

2.4. INITIATION OF OPERATIONAL MONITORING

The tasks include:

- After realistic MODIS band 2 radiance or reflectance thresholds are established for each gauging reach, implementation of a MODIS-based flood detection and warning capability depends on near real time re-calculation of the subreach ratio immediately upon receipt of new MODIS data. The new value is compared to the alarm threshold. For example, for the data displayed in Fig. 3, the bankfull alarm threshold occurs at a ratio of 1.8, and as indicated by comparisons of mapped inundation extent (not shown) to the radiance ratio.
- The same approach can be used to establish reach "hydrologic status", including low flow as well as high flow conditions, and based on the ranking (percentile) location of the newly observed values compared to the complete time series. This is, however, beyond the scope of the present paper.

Operational use of MODIS can provide several benefits to current flood forecasting and warning activities within many government agencies or ministries. For example, it is not limited by national borders. Also, river reaches that are not otherwise being measured can be monitored via this approach. However, this application is also constrained in one important way: cloud cover is commonly associated with the surface conditions leading to flooding.

The frequency of MODIS data collection helps to alleviate this constraint because temporary breaks in cloud cover provide intermittent access to surface conditions and for many flood events. However, the sensor cannot always be relied upon, as is the case for in situ gauging stations. Due to cloud cover, MODIS observation may, for particular events at particular reaches, not be possible at all. In regions such as used in our pilot studies, data can be obtained at some gauging reaches within the region nearly every day, but at particular reaches there can occur longer time periods without new data. Operational use must consider these limitations. The denser the array of gauging reaches, the better the temporal as well as spatial coverage for a region. Fortunately, a dense array is economical to support once operational. Thus, after the time series is gathered for each reach, new data collection can be automated such that 100 gauging reaches can be monitored as easily as 10.

3. Disaster response and damage assessment

The MODIS data stream can provide very useful near-real time information to disaster responders and to damage assessment. Because the sensors are always on, there is no need to schedule sensor data acquisitions, and because of the wide-area coverage, and frequent repeat, there are commonly good opportunities to map flood inundation even if intermittent cloud cover is occurring. Fig. 4 is one example of several hundred flood events so far mapped by Dartmouth Flood Observatory personnel using MODIS. Illustrated is a portion of the complete map.

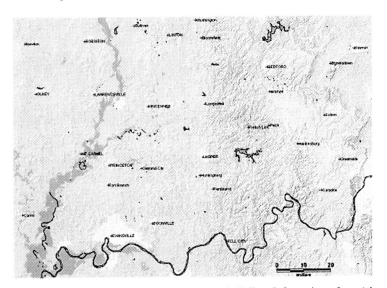

Fig. 4. Rapid response inundation map of the Wabash Valley (left portion of map) based on MODIS data. The light grey tones show areas flooded, the dark grey outline encloses water imaged by MODIS during non-flooded conditions. Shaded relief base material was generated from NASA Shuttle Radar Topography Mission (SRTM) topographic data

The wide-area nature of MODIS should not obscure the fact that it can be used for inundation mapping even at relatively large map scales. For both large and small-scale map products, a similar operational methodology is employed:

- The rectified, geocoded, bands 1 and 2 MODIS swath data are subjected to a non-supervised pixel classification algorithm, such as the isodata classifier supported by Envi™.
- One or more of the resulting classes are visually identified as water pixels, and combined into one class, if necessary.
- The same software is used to fit geographic information system (GIS) digital vector polygons around all "water" pixels.
- The inundation vectors are superimposed onto false colour composites using both bands of the same MODIS scene, and "false positive" vectors (commonly, cloud shadows) are deleted. Also, in this quality control step, areas of the map missing due to heavy cloud cover are delineated for depiction on the final map.
- The edited GIS vectors are then superimposed on map base material, and over previously mapped water, in order to produce the rapid response map (Figs. 4, 5).

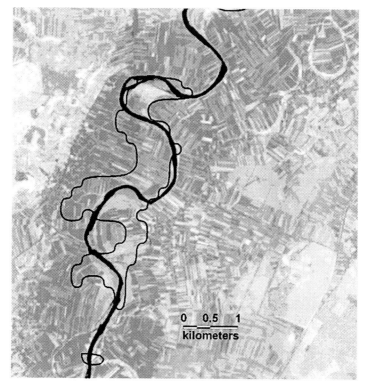

Fig. 5. Flooding along the Somes River in Romania, March 19, 2005, as recorded by MODIS (black outline). ASTER band 2 provides the base map

For large scale maps, we use NASA-provided clear sky Advanced Space borne Thermal Emission and Reflection Radiometer (ASTER) data, with a spatial resolution of 15 m. Note that the MODIS data are at much coarser spatial resolution, but still provide very useful information when superimposed over the large scale base material. This is the best approach to identify, for example, human settlements and transportation links likely to be affected by the flooding.

4. Flood disaster prevention and mitigation

Routine MODIS-based mapping of the land inundated by floods, each year since early 2001, and continuing to the present, can and should be accomplished by local, regional, and national government agencies. The motivation for such work is uncomplicated: through such work, it is possible to objectively document those lands subject to flood hazard. Fig. 6 is an example of such a MODIS-derived flood hazard map, for a transboundary area in Southeast Asia. In this case, the river itself forms the international boundary.

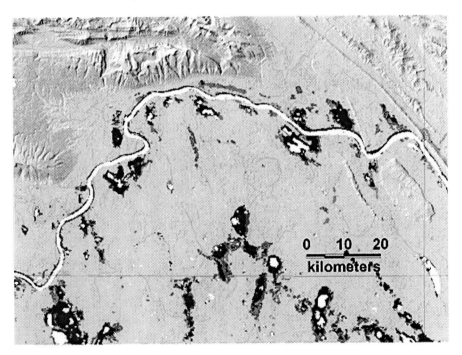

Fig. 6. Multi-year MODIS-based mapping of flood inundation, Mekong River, where it forms the boundary between Laos (north) and Thailand (south). Shaded relief base material was generated from NASA Shuttle Radar Topography Mission (SRTM) topographic data

Operational implementation of MODIS for flood hazard mapping commonly serves different user communities and is also, primarily, a GIS rather than remote sensing technical activity. The final outcome is high quality maps that are suitable for international agency (Caquard et al. 2004), government regulatory, insurance company and for other commercial users. In the example shown, flooding that occurred in different years is depicted by varying colours: maps using this cartography allow the user to determine in which year maximum flooding occurred. However, there are other methods for displaying the same information to emphasize other information. Also, these maps are not static end-products but should be updated each year, as new flooding occurs.

We emphasise the synergy between rapid response mapping, utilising the always-on MODIS sensor, and the use of GIS software, which can manipulate and store inundation information. Even as real-time inundation maps are provided for the use of government ministries and the affected populations, the data can be incorporated into GIS-based flood hazard maps. Preservation of such maps within a geographic information system provides exceptionally valuable and objective basic information for evaluating flood hazard on an international basis and also for forecasting the inundation effects of future flooding.

5. Conclusions

Three potential operational applications of MODIS-based observations of inland flooding along streams and rivers have been presented, and their implementation steps outlined. The applications include flood detection and warning, rapid response flood inundation mapping, and flood hazard assessment (by map compilation, over time, of MODIS-observed inundation).

All of these applications, but perhaps especially compilation of region-wide coverage of many regional floods, over time, would be very expensive using traditional orbital remote sensing data. Individual scenes covering relatively limited land areas must be purchased, and personnel time required for their analysis is much higher than for MODIS information. A revolution of orbital sensor capability is now underway, however, and because the MODIS 250 m spectral bands, with contained geocoding information, provide frequent, wide-area, and free coverage of water area changes. Such changes can now be observed internationally using some of the same methodologies long used by meteorologists and climatologists applied with weather satellite data.

Operational application of MODIS and the planned follow-on sensors (for example, the NASA/NOAA Visible/Infrared Imager/Radiometer Suite,

VIIRS, aboard the National Polar-orbiting Operational Environmental Satellite System, NPOESS) for hydrological purposes is in its infancy. However, the potential has been clearly demonstrated. It can be implemented on an international basis as long as continued public access to such data is maintained and internet data distribution facilities are supported.

References

Brakenridge GR, Anderson E, Nghiem SV, Caquard S, Shabaneh T (2003) Flood warnings, flood disaster assessments and flood hazard reduction: the roles of orbital remote sensing. Proc. of the 30th International Symposium on Remote Sensing of the Environment, Honolulu, Hawaii, November 10–14

Brakenridge GR, Carlos H, Anderson E (2003) Satellite gauging reaches: a strategy for MODIS-based river monitoring. 9th International Symposium on Remote Sensing, International Society for Optical Engineering (SPIE), Crete, Greece, and Proceedings of SPIE, Vol. 4886, pp 479–485

Brakenridge GR, Nghiem SV, Anderson E, Chien S (2005) Space-based measurement of river runoff. EOS, Trans. Amer. Geophys. Union, 86(19): 185–192

Caquard S, Brakenridge GR (2004) Large floods on the Kenyan side of Lake Victoria. Report and large-scale map for the Associated Programme on Flood Management, UN-World Meteorological Organization, Geneva

Marsalek J, Stancalie G, Brakenridge R, Ungureanu V, Kerenyi J, Szekeres J (2004) NATO Science for Peace project on management of transboundary floods in the Crisul-Körös river system (Romania-Hungary). In: Flood Risk Management: Hazards, Vulnerability, Mitigation Measures, Proc. of NATO Advanced Research Workshop, Ostrov u Tise, Czech Republic, Oct. 6-10, pp 191–202

Mertes LAK, Dekker AG, Brakenridge GR, Birkett CM, Le Toueneau G (2004) Rivers and lakes. In: Manual of Remote Sensing, Vol. 5, Natural Resources and Environment, Ustin S, Rencz A (eds), John Wiley and Sons, NY, pp 345–400

EXPERIENCE WITH DISCHARGE MEASUREMENTS DURING EXTREME FLOOD EVENTS

JANOS SZEKERES[1]
"VITUKI" Environmental Protection and Water Management Research Institute, Budapest, Hungary

Abstract. In Hungary the hydrological data collection is performed by twelve District Environmental Protection and Water Authorities (DEPWA) functioning under the supervision of the National Environmental Protection and Water Authority. The flow measurements are carried out according to a centrally developed time schedule. This plan is also coordinated with the neighbouring countries. In the case of extreme (low or high) flow situations, additional measurements could be organised. The flood discharge measurement series supply very important input information for the flood forecasting models. The Acoustic Doppler Current Profiler (ADCP) technique increases the accuracy of the results and the number of measurements in time and space. Last but not least, the cost of the flow measurements could be reduced, too.

Keywords: discharge measurement, ADCP technique, rating curve, discharge loop, longitudinal discharge profile

1. Organisation of hydrological data collection in Hungary

The hydrological data collection in Hungary is performed by twelve District Environmental Protection and Water Authorities and their work is supervised by the National Environmental and Water Authority. VITUKI has the

---

[1] To whom correspondence should be addressed. Janos Szekeres, VITUKI Environmental Protection and Water Management Research Institute, Kvassay Jenő út 1., Budapest, Hungary, H-1095; e-mail: Szekeres@vituki.hu

task to coordinate the measurements and the development of methods. The operating regions of the twelve DEPWAs are shown in Fig. 1.

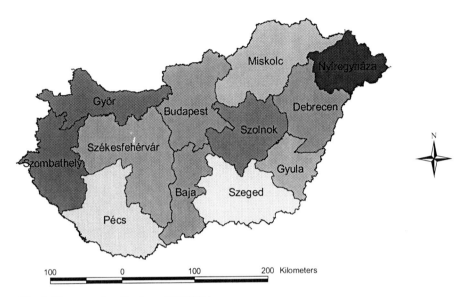

Fig. 1. The operating districts of DEPWAs

Each of them has two or three discharge measuring teams able to accomplish measurements from boat or bridge. During flood periods, the discharge is usually measured from bridge only because of dangerous circumstances, or from large safe boats, which can be anchored by the measuring verticals. The groups use mainly SEBA and OTT current meters and accessories, but the measuring boats and jib carrying carts are not standardised.

Regularly, five to twenty measurements are performed annually at each station, following a common measurement time schedule, which ensures that the measurements along larger rivers or river systems are taken simultaneously (the Danube River, the Tisza, etc.).

We have made cooperation agreements with neighbouring countries to harmonise the measurement time schedules at the border stations, as well as to create an opportunity to compare the results of discharge measurements and later on to compare the computed flow data series. Every year, two or three expert meetings are organised with the representatives of institutes of the neighbouring countries for the comparison of measured and computed results. Figure 2 shows the border stations, where regular data exchange, comparison and evaluation of measured and calculated data are carried out.

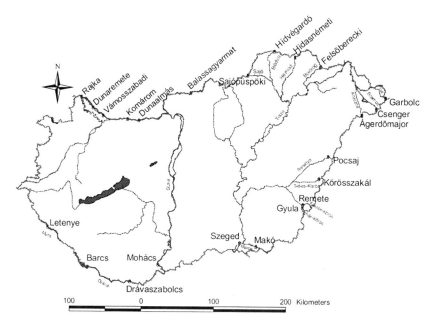

Fig. 2. Border stations

2. Measurements under extreme flow conditions

In the case of extreme low or high flow periods, additional measuring programs are organised. During very dry periods, all the rivers of the referenced catchment area are measured practically simultaneously, according to a centrally determined measuring schedule. Fig. 3 indicates sites of measuring sections on the Danube and Tisza Rivers in 2003. As a result of these measurements, we constructed the longitudinal low-flow discharge profiles of those rivers, one of which, for the Tisza River, is shown in Fig. 4.

During floods, the discharge measurements are carried out more frequently at the selected measuring sections, depending on the magnitude of the flood wave. If the gauge reading exceeds the I. Degree Warning Level (I. DWL), one measurement should be made on both the rising and the falling limbs of the flood wave.

Above the II. DWL, measurements are performed every second day, and above the III DWL, every day in the above sections.

The results of these measurements are very important input data for the forecasting models, which provide the flood defence headquarters with online information.

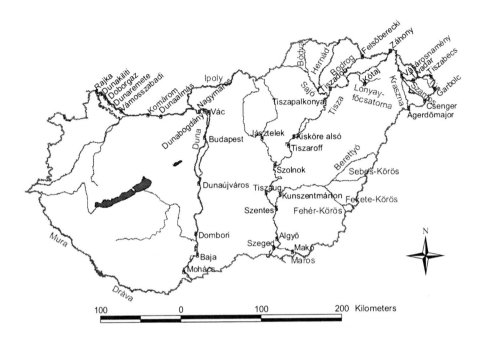

Fig. 3. Low flow measuring stations

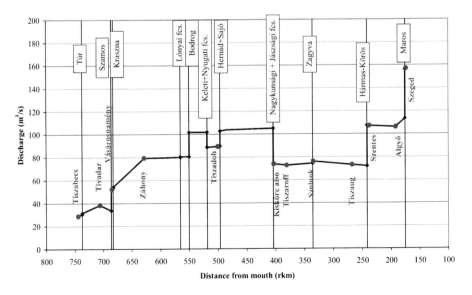

Fig. 4. Longitudinal low flow discharge profile on the Tisza River, Aug. 26-29, 2003

This means fewer measurements are done in the upper reaches of rivers because there the flood waves are usually rapid and flashy, but in the middle or lower reaches, the durations of floods are longer due to the significantly smaller surface slopes. As a result of this, for example in 1970 on the Maros River at Makó the flood discharge measurements started on April 7 and finished on May 19. During this period, 39 measurements were accomplished. There were three separate bridges (one on the main channel, two on the left and right flood plains), with 16 openings and the total number of measuring verticals were 40. In the vicinity of the peak flow, every second day suspended sediment measurements were also carried out in the section (16 in total).

Since the high water periods on the Danube River and the Tisza River usually do not coincide, the measuring groups are transferred from the unaffected Water Authority to the one experiencing flooding. Figure 5 presents the measuring sections on the Tisza River and its tributaries during the 2000 flood. The number of flood measurements performed is also indicated together with the name of the corresponding station.

Stations	Number of measurements	Time period
Tivadar	16	10.03-13.04
Csenger	13	10.03-13.04
Vásárosnamény	27	11.03-16.04
Záhony	23	11.03-16.04
Felsőberecki	18	29.03-16.04
Tokaj	18	29.03-16.04
Kisköre	79	08.02-06.05
Szolnok	40	16.03-11.05
Vezseny	21	10.04-04.05
Tiszaug	13	22.04-04.05
Csongrád	14	17.04-04.05
Szentes	34	07.04-15.05
Algyő	36	07.02-15.05
Szeged	34	08.04-16.05
Makó	28	13.03-14.05

Fig. 5. Flood discharge measuring stations on the Tisza River in 2000

As a result of the long series of measurements, the so-called loop rating curves may be studied at several stations. This is rather characteristic in the flatland regions of Hungary and is caused mainly by the moderate water surface slopes.

Figures 6-8 show the results of the discharge measuring series in different years on the Tisza and Maros Rivers. In Fig. 6 the loop rating curves measured at Szolnok in 1970 and in the years 1998-2001 are presented.

It can be noted that the maximum water levels in 1970 and 1998 were nearly the same, but the peak flow in 1998 was approximately 300 m³/s smaller than in 1970. In 1999 the peak flow was nearly the same but the culmination water level was higher by 70 cm. The cause of this is that the flow conveyance capacity of the river bed in this region has deteriorated since 1970. The situation was similar in 2000, but in 2001, the flood plain was cleared of the dense vegetation before the flood, and the discharge loop has shifted back to its original (1970) position.

The discharge loops show significant differences between the discharges corresponding to the same stage on the rising and falling limbs of the flood hydrograph. For example, at 900 cm the two discharges are 2,540 and 1,890 m³/s, respectively. The 650 m³/s difference is 30% of the average value 2,215. This is the cause, why in the middle and lower reaches of River Tisza the simple two-segment rating curves are not useable.

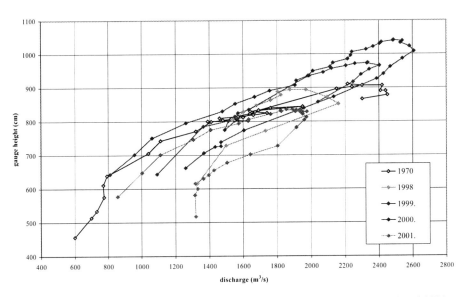

Fig. 6. Loop rating curves on the Tisza River at Szolnok in 1970, 1998, 1999, 2000 and 2001

In Fig. 7 the results of measurements performed on the Tisza River at Kisköre in the years 1998-2001 are plotted. The discharge loops show significant differences among the discharges corresponding to the same gauge stage on the rising and falling limbs of the flood waves.

The results of discharge measurements carried out during a flood wave having four peaks in 1970, on the Maros River at Makó, are presented in Fig. 8, showing the multiple discharge loops together with the discharge time series.

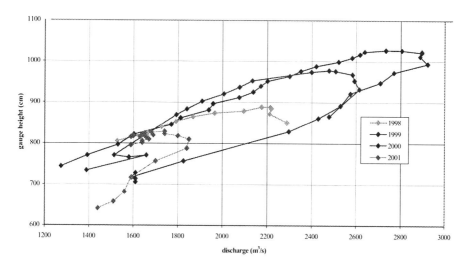

Fig. 7. Loop rating curves on the Tisza River at Kisköre in 1998, 1999, 2000 and 2001

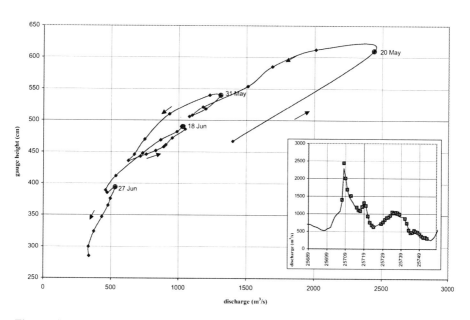

Fig. 8. Discharge loop on the Maros River at Makó, 1970

3. Use of the ADCP in flood discharge measurements

All the discharge measurements were carried out applying the "velocity-area" method, which is known to require considerable time and manpower.

In the case of floods the discharge measurements in bridge sections may require two or more teams working simultaneously; one in the main channel and the other (or others) measuring in the flood plains. The complete measurement may take several hours, during which the hydraulic conditions might change significantly (rapidly rising or falling water levels). Thus, the final result of the measurement does not correspond to an exact point in time and stage, but rather to the mean of a longer time period and the corresponding average gauge stage. In addition, the accuracy of measurements may be adversely influenced by the disturbed flow pattern in the downstream section of bridge openings caused by the piers, eddies and by the widely varying flow directions.

In August of 2002 an extreme flood wave has arrived on the Danube River to Hungary. At this time our ADCP (Acoustic Doppler Current Profiler) equipment was being tested in the field. As a result of this, simultaneously with the ADCP measurements, the traditional velocity-area method using propeller-type current meters was applied, too. Only one current-meter measurement was accomplished at the Nagymaros section, using a large ship with a four-person measuring group and a three-person navigation crew on board. During the same time, the ADCP measurements were performed using a simple motorboat, operated by two persons (one boat driver and one ADCP operator) in four sections, with four repetitions in each, between Nagymaros and Budapest. This time included the transportation of the motor-boat from Budapest to Nagymaros (40 km) and travelling back on the river downstream between the measuring sections.

The results of the traditional and the ADCP measurements compared well and proved the accuracy of the new method. The data obtained by the four repeated measurements agreed with each other within ±2%, and the differences between the results of the current meter and ADCP measurements did not exceed 5%. This was a convincing example of the efficiency of the ADCP equipment as an instrument for rapid, cheap and accurate collection of data.

The only significant problem observed with ADCP measurements was caused by the mobility of the alluvial river beds. For example, within the City of Budapest, where the width of the riverbed was confined to 350 m and the flow velocity was higher than 3 m/s, the measurement in the middle of the cross-section was impossible due to the significant bed-load transport.

The results of the discharge measurements are presented in Fig. 9 and show the change of hydrographs of the flood wave at the various stations along the river (their locations are indicated in the legend).

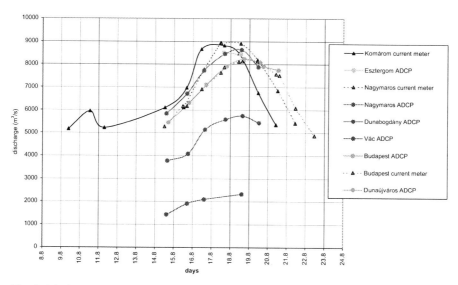

Fig. 9. Discharge measurements on the Danube River between Komárom-Dunaújváros in August of 2002

The Dunabogdány and Vác stations are situated on two different arms of the Danube River between Nagymaros and Budapest. Consequently, the sum of the two corresponding discharges should be compared with the nearest upstream and downstream discharges measured in the unified river channel.

4. Conclusions

The flow rate measurements produce very important results for water management, river training, flood protection, navigation, etc. Especially, the extreme values are very informative, and consequently, the collection of such data is one of the most important tasks of hydrologists. The missed measurements are not replaceable. The discharge data are of particular interest in international rivers. Hungary is situated in the Carpathian basin. The rivers enter here from six neighbouring countries, so the international hydrological collaboration has to be well organised and operated. Most of the territory of Hungary is flatland. The composite channels of regulated rivers are bordered by flood protecting dykes, keeping the flood hazards away from municipalities, agricultural and industrial areas. Consequently, the flood forecasting requires most up to date information about the flow conditions. The flood discharge measurements have a long tradition in Hungary. This hard and sometimes dangerous work has recently been significantly improved by a new instrument, the ADCP, which contributes to the higher accuracy, efficiency and speed of discharge measurements, while reducing the operational costs, and the last but not the least, increasing the frequency of measurements in time and space.

DEVELOPMENT OF THE HYDROMETEOROLOGICAL TELEMETRY SYSTEM IN THE KÖRÖS RIVER BASIN

ATTILA KISS[1], BÉLA LUKÁCS
*Körös Region Environmental and Water Directorate,
Városház u. 26. H-5700 Gyula, Hungary*

Abstract. The Körös river system is rather complicated, with frequent rapid floods and long lasting low-water periods. The catchment area is divided between two countries (Hungary and Romania) and, therefore, collaboration between the water management agencies of both countries is essential. One of the best tools for maintaining good data transmission and exchange is the up-to-date telemetry system requiring continuing development. The following contribution describes the completed stages of the development of the telemetry system for hydrometeorological data on the Hungarian side and suggests further improvements and future developments of the system.

Keywords: telemetry system, floods, hydrology, hydrometeorology, remote sensing, data transfer

1. Introduction

The need for the Körös hydrometeorological telemetry system follows from the specific features of the Körös drainage basin and the current local water management activities.

---

[1] To whom correspondence should be addressed. Kiss Attila, Körös-Vidéki Környezetvédelmi és Vízügyi Igazgatóság H-5700 Gyula, Városház u. 26. Hungary; e-mail: kiss.attila@korkovizig.hu

1.1. FEATURES OF THE CATCHMENT AREA

The entire catchment area of the Körös rivers is 27,537 km^2, and more than one half of this area is on the territory of Romania. The annual precipitation varies from 500 mm in the lowlands on the Hungarian side up to 1200 mm in the mountainous areas on the Romanian side. The water is collected and drained by four major tributaries (Fehér-, Fekete-, Sebes Körös and Berettyó), with complex interactions and backwater effects on the Hungarian side. The lower reaches of the Hármas-Körös River (flowing into the Tisza River) are directly affected by water levels in the Tisza. The water level of the rivers is not uniform; the distribution of their inflows is extreme. Floods are heavy, but there is often a period of water shortage characterised by increased sensitivity of water quality conditions. Most of the region is a low flood plain, with 69% of the managed territory representing an area protected by dykes and serving for agriculture and food industry. Both these sectors are particularly sensitive with respect to water damage prevention. Drainage of high excess surface water flows and high density of structural build-up characterise the controlled territory. On behalf of the state administration, the Körös Region Environmental and Water Directorate manages field activities on an area of 4,108 km^2.

During large floods in the Körös valley in the years of 1966, 1970, 1974, 1980, 1981, 1995 and 2000, the highest water levels exceeded several times the maximum values observed before, and also caused dam breaks and flooding, and necessitated the construction of detention reservoirs. In the Körös rivers valley, the construction of detention reservoirs started following the 1970 flood: the Mályvádi reservoir on the Fekete Körös was built in 1975-79, and the Mérgesi reservoir at the delta of the Kettős-Körös and the Sebes-Körös in 1975. On the right bank of the Fehér-Körös, the Kisdelta impounding reservoir was built in 1997-98. With their combined temporary water storing capacity of 188 million m^3, these reservoirs enable critical reduction of flood levels.

1.2. HYDROGRAPHIC FACTORS

The hydrologic service organisation of the local water management sector operates a monitoring network to provide the basic hydrologic data needed for water management activities. The data from the 129 national stations on the territory are supplemented by data from more than 500 additional stations serving for operational needs or study purposes.

2. The needs for the telemetry system

The most important reasons for implementing a hydrometeorological telemetry system are the following:

- Increasing the lead time on torrential streams, to prevent or reduce flood damages,
- Serving locations, where data collection is most difficult and unsuitable for deploying human observers,
- Improving accuracy of forecasting models,
- Providing continuous data series where needed for operational reasons (e.g. operation of detention reservoirs),
- Providing data for operation of control tasks (e.g. automatic directing systems, operation of navigation locks, etc.),
- To reduce labour costs (replacing human observers), and
- Executing complex measurements.

All the above mentioned points (except the second one) apply in the case of the telemetry system established on the territory of the Directorate; after all, the most important requirement of the flood defence management is to learn about floods as soon as possible and with the greatest possible accuracy. Because of the short lead times of flood peaks, measures must be taken for the preparation of many defence works. These measures have large financial consequences in certain cases, because they determine the manpower and technical support that has to be mobilised, making decisions whether to activate emergency detention reservoirs, and whether to evacuate communities for the sake of safety, etc.

At present, we can use only multivariable regression models for flood forecasting, but our long-term plan is to use hydraulic models in the entire Körös system and such models need much more information about the water-courses. Application of a joint telemetry system with the Romanian colleagues would be a perfect solution to make accurate flood forecasts by means of hydraulic models. A NATO SfP Project started in 2002 to develop such a program.

A good example of needs for continuous data series is the station of Sebes-Körös at Körösszakál. Upstream from this station, on the Romanian side, a series of dam reservoirs is operated to produce electric energy. Therefore, each day the water level at this station fluctuates between low and high levels as shown in Figure 1.

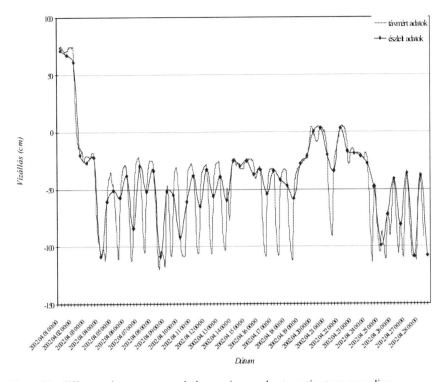

Fig. 1. The difference between manual observations and automatic stage recording

In the figure the dotted line shows the data reported by the observer twice a day, as done in the traditional measuring system, while the continuous line shows the data recorded continuously in time. As it can be seen much of the important information is missing from the manual measurements, because of the operation of reservoirs. If only manual measurements were done at this border station, the determination of the actual daily flows would be very inaccurate.

In the Körös region, one of the most important operational control tasks is to operate the dam reservoirs during the low-water season. This problem can be handled with the telemetry system, because there are two telemetry stations at Békésszentandrás and Békés, which are the most important elements of the water management system of the Körös rivers.

After the political and social changes of the last decade in Hungary the human manpower became very expensive. These automatic stations make the measurements possible at those places where human manpower is not available to maintain continuous data series collection at hydrometric stations. Since the beginning of 2003, we are working under the ISO

9001:2001 quality control system and the use of the telemetry system is an optimal solution satisfying the ISO requirements.

Also a great advantage of a telemetry system is that it is possible to measure many more hydraulic or water quality parameters at the same place and in the same time. Presently we are extending the measurements at certain sites for the measurements of water temperature, flow rate and precipitation or air temperature at some stations. The future plan is to enlarge this spectrum mainly for water quality parameters.

2.1. STATISTICS OF THE TELEMETRY SYSTEM

The first complex telemetry system in Hungary was established in West-Transdanubia on the Zala river, which remains in operation with the help of continuous modernisation until now. Currently, telemetry systems are operated in Hungary on all the major watercourses.

The first telemetry stations in the Körös region were established in 1999, and presently, there are 20 stations in operation. The system stations are listed in Table 1.

2.2. FINANCIAL ASPECTS

Besides the Directorate's own financial resources, the contribution of the "Joint Flood Monitoring on the River Fehér-Körös" PHARE CBC project (1997-99), under reg.no.:ZZ9622-03-03, has been of a major importance in the development of the system. In total, 12 stations were installed under this project with a total PHARE CBC contribution of 90,000 Euro.

The telemetry system on the territory of the Directorate can produce adequate information continuously, including the periods of hydrological extremes, which occur in the Körös region. The telemetry system is able to provide accurate data for both the very rapid floods as well as the low flows with backwater effects caused by dams during summers. It should be noted that during the system development, since the early 1970s, most stations with data recorders already had stilling wells. Therefore the station construction did not incur extra costs during the development of the new telemetry system.

Table 1. Stations in the Körös region telemetry system

	Name of the watercourse	Name of the station	Measured elements
1	Fehér-Körös	Gyulai duzzasztó felső	Water level, discharge
2	Fehér-Körös	Gyula	Water level, water temperature
3	Fekete-Körös	Ant	Water level, water temperature
4	Fekete-Körös	Sarkad-Malomfok	Water level, discharge, water temperature
5	Fekete-Körös	Remete	Water level
6	Kettős-Körös	Doboz	Water level
7	Kettős-Körös	Békési duzzasztó felső	Water level, water temperature, dam-gate position
8	Kettős-Körös	Békési duzzasztó also	Water level
9	Kettős-Körös	Békés	Water level
10	Kettős-Körös	Mezőberény-Hosszúfok	Water level, water temperature
11	Kettős-Körös	Köröstarcsa	Water level, water temperature
12	Sebes-Körös	Körösszakál	Water level, water temperature
13	Sebes-Körös	Körösladány	Water level
14	Berettyó [1]	Szeghalom	Water level, water temperature, precipitation, air temperature
15	Hármas-Körös	Gyoma	Water level
16	Hármas-Körös	Kisőrvető	Water level
17	Hármas-Körös	Szarvas	Water level, water temperature
18	Hármas-Körös	Békésszentandrási duzzasztó felső	Water level, dam-gate position
19	Hármas-Körös	Békésszentandrási duzzasztó also	Water level
20	Hármas-Körös	Kunszentmárton	Water level, water temperature

[1] Downstream from the national border in Berettyó there are two stations (Pocsaj and Berettyóújfalu) with telemetry installations operated by the Trans-Tisza Environmental and Water Directorate.

2.3. THE STRUCTURE OF THE TELEMETRY SYSTEM

The telemetry system has two main units. One is the centre of the system which is an up-to-date PC located in the Hydrological Group offices at the Directorate headquarters. The PC has system control software ESZTER, which was developed by the Hungarian Restart Plc company, and receives the data sent by the stations. The software program can facilitate immediate data retrieval, when the selected station receives an order to send instantly to the centre the data measured by the water-level sensors. This function makes data retrieval possible at all stations and at selected times.

A very important part of the program is the possibility to retrieve the data stored in the station datalogger. The stored data originate at the telemetry station every 30 minutes, as requested and stored by the program according to the preset time period. Using these requested and received water-level data the central computer can compare them to a preset alarm threshold value. If the received water level is higher than the threshold value, the program will produce an alarm signal on the screen immediately. There is also a possibility to redirect the alarm signal to a preset telephone number. If the water-level is higher than the preset threshold value, the measuring frequency automatically switches to a more frequent mode. The program also facilitates the display of the stored data in a table or in a graphic format. The final processing step executed by the ESZTER program is to send the data to the SQL server. The main screen of the telemetry system as shown in the centre is displayed in Figure 2.

By clicking on any of the stations in Fig. 3, the pertinent information is shown. As shown in Fig. 3, all current hydrometeorological data and service information are available not only in numeric, but also in graphic formats as well. Also, certain settings can be changed by commands issued at the centre, using the control program.

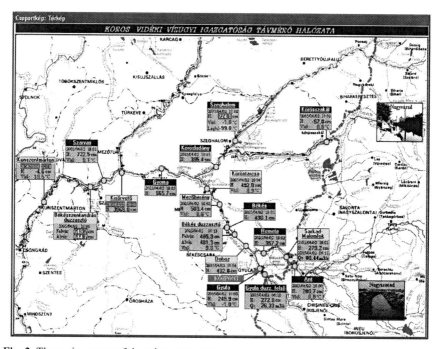

Fig. 2. The main screen of the telemetry system

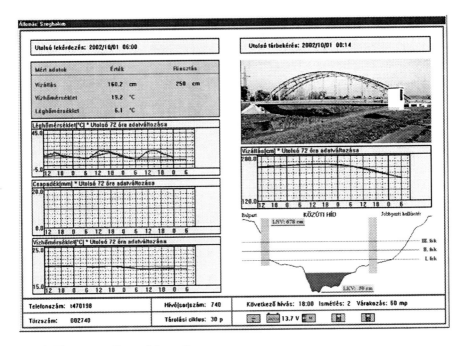

Fig. 3. The screen of one of the stations

The second main unit of the telemetry system is the network of the observation stations, representing telemetry units located along the river system. These units consist of two parts.

2.3.1. *Telemetry unit*

Telemetry units include the following equipment:
- A data recorder well, with float water-level sensors (OTT OWK-16 type, accuracy of 1 mm), water temperature sensors, ultrasonic discharge measuring equipment (OTT Sonicflow type), signal converters, and intruder sensors
- A dyke (levee) guard house containing Electronic data loggers (Gealog S type), Communicators (RKOG type), Signal converters, Telephone modems, Continuous electric supply units, Water level displays and Intruder sensors.

2.3.2. *Directorate centre*

At the Directorate Centre, the equipment includes: telephone modems, an Uninterrupted Power Supply unit, computer system, and the ESZTER type system-controlling software.

3. Planned future developments

Two ways can be foreseen for the extension of the existing telemetry system. The system can be further developed quantitatively or qualitatively. The quantitative development involves installation of additional stations. However, this type of development has been already stopped, because the main water-courses in the Körös region are already well covered by the existing stations. A qualitative development involves improving the measurement of data and data transmission techniques using the most up-to-date instruments and also extending the list of the measured elements.

The idea of connecting the existing telemetry systems between the Hungarian and Rumanian part of the basin remains an important issue, which is now being studied by the Hungarian-Rumanian Water and Technical Joint Committee.

A very informative website already exists for the Körös Valley Environmental and Water Directorate, but the data from the telemetry system are still available only in the local area network of the Directorate and to those Hungarian users who have special software that can be used with SQL server's data. One of the goals is to put the data from the telemetry system onto the Internet.

EXPERIENCE FROM OPERATION OF THE JOINT HUNGARIAN-UKRAINIAN HYDROLOGICAL TELEMETRY SYSTEM OF THE UPPER TISZA

KÁROLY KONECSNY [1]
"VITUKI" Environmental Protection and Water Management Research Institute, Budapest, Hungary

Abstract. Owing to the torrential character of the Upper Tisza and its tributaries, an efficient flood forecasting and early warning system has to be based on telemetry. Following the development of the joint Hungarian-Ukrainian hydrological telemetry system in 2003, the system now provides reliable performance. Further developments will aim to connect the operating and planned systems of both countries in the transboundary region and establish a joint monitoring system.

Keywords: telemetry, hydrological telemetry system, sensors, data loggers, retranslation station, telemetry centre, uniform monitoring system

1. Introduction

Floods of the period 1998-2000 proved that significant floods may occur in any season on the Upper Tisza and tributaries. Steeply rising limbs of flood waves may lead to high water levels requiring intensive protective actions within 24 hours at the transboundary cross-section between Hungary and

---

[1] To whom correspondence should be addressed. *Károly Konecsny*, „VITUKI" Environmental Protection and Water Management Research Institute, Kvassay Jenő út 1., Budapest, Hungary, H-1095, e-mail: konecsny.k@vituki.hu
Formerly at the Upper Tisza Environmental and Water Directorate, Nyíregyháza

Ukraine. These hydrological conditions define the possible solutions for flood protection and the requirements on data collection, processing and storage. The joint Hungarian-Ukrainian hydrological telemetry system aims to achieve rapid and reliable data collection and supply of good quality information for improving the efficiency of decision making in water management and flood defence. In the latter case, increasing the lead time and accuracy of hydrological forecasting is particularly important.

A historical review was given in earlier studies (Illés and Domokos 1989; Illés and Konecsny 1995; Illés et al. 1999; Konecsny 2000, 2003, 2005; Illés 2002; Pasche et al. 2002) describing the course of the establishment of the Upper Tisza hydrological telemetry system. An important stage of the system development in the Ukrainian part of the basin was completed with the formal launching of the telemetry network on Oct. 30, 2003, in the presence of supervising Ministers of the two countries. Since that time, the system has been operated without interruption and a great volume of hydrological, meteorological and system operation data has been collected. Analysis of technical data of the system operation, and good and bad experiences gained, should be useful for designing, constructing and operating new telemetry systems in other catchments.

2. Development of the hydrological telemetry system in space and time

Four different generations of telemetry technology followed one another in the course of 35 years on the territory of the Upper Tisza Environmental and Water Directorate (FETIKOVIZIG). These systems reflecting technological informatics and data transmission possibilities of their given eras were the following ones: 1968-78 TELEXDAT, 1978-88 HYDRA, 1988-98 MIKI M80, and since 1998, INTELLUTION-ELCOM. Reliability of the automatic units has been gradually improved, but for a long time it has not reached the level required for uninterrupted flood observations. Operational reliability of 75-80% has been reached on the average. Owing to their inadequacies these generations of telemetry systems could not meet the set requirements; however, they produced practical experience, which could have not been gained without their operation.

The following key requirements were defined before the most recent development started (Illés 1995):

- Reliability of data transmission should be kept above 95% and the accuracy of data should allow less frequent manual observations.
- The system should be composed of universal elements.
- Commercially available hardware and software tools should be used.

- Extensions and new versions of the system should not be restricted to the original contractor.
- Recommendations from the national network and their applicability in the Upper Tisza region should be considered.
- The telemetry system should be linked to the computer network of the Directorate; data from the regional telemetry centre should be fed into the national information system and also made available to the engineering districts.
- Usable elements of existing telemetry systems should be utilised.
- The development of the system should be harmonised with the plans for the Trans-Carpathian region in Ukraine.

The upgrading of the system composed of 10 units (centre, re-transmission station, and hydrological stations) was completed in 1998. The following sensors were put into operation: water level, precipitation, air temperature, water temperature, transmission error, and indication of illegal intrusion or theft.

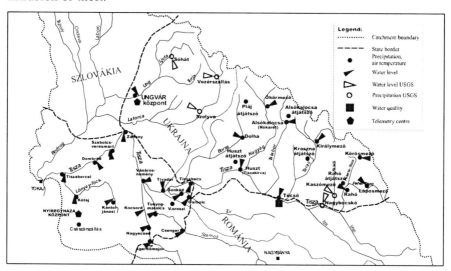

Fig. 1. Scheme of the joint Hungarian-Ukrainian hydrological telemetry system

100 million HUF were granted in 2000 under the Hungarian government programme for the development of the system in the Trans-Carpathian region of Ukraine. This system is compatible with the Hungarian system and maintains wireless and microwave links. Centres in Uzhgorod and Nyíregyháza were linked with a wide band microwave channel and data exchange was performed with 5-minute frequency. The system contained 11 units at that time in the Trans-Carpatian region:

Uzhgorod centre, 8 wireless retranslation stations, and 2 precipitation and air temperature stations (Illés 2002; Konecsny 2003). This network did not meet the hydrological requirements with respect to the station density. Following the flood event in March 2001 additional 215 million HUF aid was given in 2003 by the Hungarian government to Ukraine to complete additional 8 hydrological stations. Four precipitation and air temperature stations were installed at the hilltop retranslation station (Konecsny 2004). Together with the two stations constructed in 2000, the total number of stations in the network increased to 14.

Table 1. Main characteristics of the stations in the hydrological telemetry system

Stream/ Catchment	Station	Distance from the mouth (km)	Elevation (m)	Range of water level changes (m)	Year of installation	Elements measured[1]	Data transmission
Tisza	Tiszabecs	744.3	114.34	9.98	1998	HPTT$_v$	WIRELESS
Tisza	Tivadar	705.7	105.40	13.07	1999	HPT	WIRELESS
Tisza	Vás.namény	684.4	101.94	11.67	1998	HPT	WIRELESS
Tisza	Záhony	627.8	98.14	10.80	1998	HPT	WIRELESS
Tisza	Dombrád	593.1	94.06	10.96	1999	HPT	WIRELESS
Tisza	Tiszabercel	569.0	91.36	10.29	2002	HPT	WIRELESS
Túr	Garbolc	27.7	116.50	7.91	1998	HPT	WIRELESS
Túr	Sonkád	11.5	112.59	6.29	2002	HPT	WIRELESS
Szamos	Csenger	49.4	113.56	10.05	1998	HPTT$_v$	WIRELESS
Szamos	Tuny.matolcs	21.9	106.21	10.67	1998	HPT	WIRELESS
Kraszna	Ágerdőmajor	44.9	110.39	6.77	1998	HPTT$_v$	WIRELESS
Kraszna	Nagyecsed	27.3	107.76	7.00	2002	H	WIRELESS
Kraszna	Kocsord	22.6	106.65	7.4	2002	HPT	WIRELESS
Túr-belvíz	Sonkád	0.2	111.97	1.50	2002	HPT	WIRELESS
Lápi fcs.	Nagyecsed	0.0	108.07	3.10	2002	HPT	WIRELESS
Belfő-cs.	Tiszabercel	0.0	91.36	1.95	2002	HPT	WIRELESS
III. sz. ff.	Kántorjánosi	32.7	139.11	1.47	2002	HPT	WIRELESS
Lónyay-f.	Kótaj	21.2	90.58	4.38	2000	HPT	WIRELESS
Rétközi-tó	Sz.veresmart	-	96.80	3.92	1998	HPT	WIRELESS
Tisza-Sz.	Vámosoroszi	-	113.00	-	2001	PT	WIRELESS
Tisza-Sza.	Kispalád	-	118.67	1.72	2004	G$_w$	GSM
Nyírség	Császárszállás	-	116.00	-	2001	PT	WIRELESS
Tisza	Rakhiv	962.0	429.73	4.55	2003	HPT	WIRELESS
Tisza	Tiachiv	887.0	208.97	8.04	2000	HPT	WIRELESS

Tisza	Khust	852.0	155.53	5.71	2003	HPT	WIRELESS
Black-Tisza	Iasinia	27.0	602.05	3.22	2003	HPT	WIRELESS
White-Tisza	Luhi	15.0	646.50	2.61	2003	HPT	WIRELESS
Tereshva	Usti Tshorna	54.0	523.86	4.78	2003	HPT	WIRELESS
Tereblia	Kolochava	56.0	524.48	2.22	2003	HPT	WIRELESS
Rika	Mizhgiria	64.0	434.22	3.88	2003	HPT	WIRELESS
Borshava	Dovge	69.0	168.35	4.29	2003	HPT	WIRELESS
Tisza	Rakhiv-Terentin retr. station	-	1369.0	-	2003	PT	WIRELESS
Tereshva	Krasna retr. st.	-	1302.0	-	2003	PT	WIRELESS
Tereblia	Kolochava retr. st.	-	1240.0	-	2003	PT	WIRELESS
Tisza	Hust Rokosovo retr. st.	-	819.0	-	2000	PT	WIRELESS
Borzhava-Lator.	Plai meteo. retr. st.	-	1339.0	-	2003	PT	WIRELESS
Ung	Uzhgorod	29.5	112.38	5.18	2003	HPT	WIRELESS
Tisza	Veliki Bichkov	945.0	294.78	5.85	2004	HPTT$_v$	Satellite
Kasovo	Kosovska Poliana	8.0	406.77	3.00	2004	HPTT$_v$	Satellite
Latoricza	Nizhnii Vorota	167.0	356.54	3.47	2003	HPTT$_v$	Satellite
Latoricza	Svaliava	138.0	190.00	4.50	2003	HPTT$_v$	Satellite
Lyuta	Chornogolovo	14.0	255.09	2.85	2004	HPTT$_v$	Satellite

[1] Remarks: H-Water level (cm), P-Precipitation (mm), T-Air temperature, T$_v$-Water temperature, G$_w$-Ground water level

The development of the system included the installation of a water quality station in Tiachiv. A station was installed on the Uzh stream at Uzhgorod with the use of a German grant. Five stations were installed in 2003-2004 in the frame of an USAID-USGS project.

Stations of the Hungarian and Ukrainian hydrological telemetry systems in operation in 2004 are listed in Table 1 and shown in Figure 1. The system can be subdivided into four functional units (Figure 2):

- Hydrological telemetry stations,
- Communication network – 450 MHz wireless and microwave links,
- Centres in Nyíregyháza and Uzhgorod – process management program,
- Connected networks – engineer workstations, intranet, national hydrological data handling systems, and a web site.

The following parameters are measured at hydrological stations and the water quality station: water level, precipitation, air temperature, and water temperature. The following types of sensors are utilised: pressure sensors, water level float gauges, radar water level sensors, water temperature sensors, tipping bucket raingauges, and air temperature sensors. Basic elements of the automated system on the Hungarian and Ukrainian side are in most cases identical. The density of the hydrological telemetry system on the territory of the FETIKOVIZIG is 1 station/248 km^2; in the Trans-Carpathian region this density is 1 station/644 km^2.

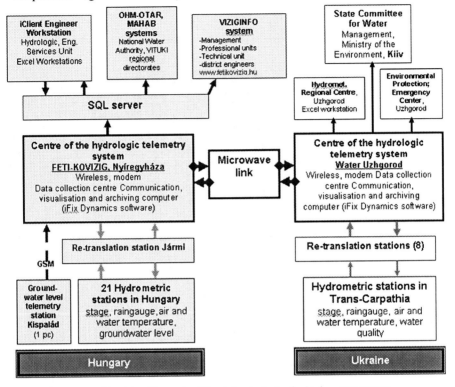

Fig. 2. Structure of the joint Hungarian-Ukrainian hydrological telemetry system

Additional 50 stations would be necessary to reach the optimal density in Trans-Carpathia. As a consequence of recent developments, 42 stations were operated at the end of 2004 in the Upper Tisza region, including 20 stations in Trans-Carpathia (see Table 1, Figure 1).

3. Operational experience during the period October 2003-October 2004

The joint Hungarian-Ukrainian hydrological telemetry system performed well during the study period (1/10/2003 – 31/10/2004) in the Upper Tisza region. The volume of hydrological information collected in the upper part of the catchment has increased by several orders of magnitude, and the time required for observations and data transmission has been significantly shortened. The system worked virtually without interruptions. Only very short stoppages occurred when the whole Ukrainian or Hungarian system was halted by failures of the central computer, software errors, power cuts, and failures of the wireless system. The duration of the stoppages with few exceptions remained within 1-2 hours. All functions of the system performed well including measurement, data collection, data transmission, data processing, visualisation and archiving. Twenty-nine stations with water level gauges produced 8,613 morning measurements at 6 a.m., and out of those 1,342 (15%) reached the Nyíregyháza telemetry centre. The largest number of missing data points (207) occurred in January 2004 and was caused by the failure of two retranslation stations in the Trans-Carpathian region. Significant improvement was observed following repairs of the transmission stations Plai and Rakhiv-Terentin in April 2004. Eighteen (2%) missing data points were noted in May 2004, which was the lowest monthly value. The percentage of missing data remained within 4-10% in the following months. Days with missing data were defined relative to the execution of the morning observations. Missing morning values were recorded as missing telemetry data for the given day regardless of the data availability during the rest of the day.

Beside the failure of retransmission stations the following other causes occurred: power cuts, failures of the pressure sensors, damaged cable insulation, water seepage into the antennas, and dead batteries. A station powered by solar energy required once the change of the solar panel. A couple of hours stoppage in the centre of FETIKOVIZIG resulted from the damage caused by lightning in October 2004.

Investigations of the spatial distribution of missing data in Figure 3 show the maximum in the vicinity of the Rakhiv-Terentin retransmission station the White-Tisza Luhi (266 – 67%), the Black-Tisza Iasinia (184 – 46%), the Tisza Rakhiv (192 – 48%); in the Hungarian part of the catchment the largest number of missing data were observed at the Tisza – Tivadar (153 – 38%), and the Tisza – Tiszabecs (116 – 29%). No data were missing at 10 stations.

Water levels reported by the telemetry system were checked and compared with manual observations on a daily basis. During flood events this comparison was more frequent. In some cases the difference between the two types of observations resulted from the mistake made by the observer. Large differences between different types of observed data occurred due to the following factors:

- The telemetry station sensor and the staff gauge were at different locations:
    - in the case of high slope, water levels are different
    - during floods, the sensor measures the water level in the main channel while the staff gauge measures that in the flood plain/flood berm.

- Error in design or construction of the stilling well, movement of the well base and silting up of the connecting tube may block water inflow. The measured water levels follow the water level of the stream with a certain delay. Water levels remain constant during prolonged low water periods.
- Calibration error of the pressure sensor. In the case of a large range of water level fluctuations (5-12 m) sudden rise or quick drop of water levels result in errors of several decimetres.
- In float gauges, the float may get stuck on the wall of the stilling well or in the silt, and other problems are caused by deformation of the cable or steel ribbon, or by freezing of mechanical elements, which may halt or slow down the movement of the float.
- Erroneous staff gauge readings.
- Errors in the position of the staff gauge, reporting values without gauge correction, or corrected by an outdated correction function.

Errors exceeding 10 cm usually occurred at stations with cross-sections with a large range of water level fluctuations. In one case the errors resulted from the elevated position of the stilling well connecting tube not working during a low flow period. From time to time, it is necessary to adjust telemetry data according to the baseline observations.

4. The methodology for the evaluation of the operation of water level telemetry stations

Mean values are misleading in the case of positive and negative errors. A monthly mean close to zero may refer only to the equivalence of positive and negative errors. Owing to this fact absolute values of errors were analysed, including monthly maxima, the number of cases exceeding 10 cm, and RMSE (root-mean-square-error).

Objective evaluation criteria were defined with the use of weighing factors. The largest weight 50 was assigned to the data transmission failure. The accuracy and reliability of water level observations were evaluated by four characteristics of the difference between the telemetry and staff gauge data (mean error, maximum of errors, days with error larger than 10 cm, RMSE), the total weight of the four parameters sums up to 50 (Table 2).

The station working without interruptions and supplying always accurate data receives the maximum of 100 points. Stations with more than 60% of missing data, a greater than 10 cm mean error, the maximum error reaching 100 cm, errors of more than 10 cm occurring longer than 60 days, and RMSE greater than 10 cm receive the score of 0. Telemetry stations in the range 71-100 are considered excellent, 51-70 good, 31-50 satisfactory, and below 30- poor.

Table 2. Score system for the evaluation of the operation of hydrological telemetry stations

Reliability of data transmission		Difference between gauge readings and telemetred data								
Missing water level data %	Score	Mean error		Maximum error		Error longer than 10 cm		RMSE		Total score – water level
		cm	score	cm	score	days	score	cm	Score	
0	50	0-1.0	10	0-10	15	0-5	15	0-1.0	10	50
0-15	40	1.1-2	7	11-20	10	6-10	10	1.1-2	8	35
15-30	30	2.1-5	5	21-40	7	11-20	7	2.1-4	5	24
30-45	20	5.1-7	3	41-70	5	21-40	5	4.1-7	3	16
45-60	10	7.1-10	1	71-100	2	41-60	3	7.1-10	2	8
>60	0	>10	0	>100	0	>60	0	>10	0	0

Based on the given system of scores, the following data were calculated for the stations investigated during the period Oct 2003 – Oct 2004: Out of the total of 28 stations, 15 stations were excellent (54%), 7 stations were good (25%), 4 were satisfactory (14%) and 2 stations were poor (7%) (Table 3). Thus, from the point of view of water level measurements more than two thirds of the stations were rated as excellent and/or good.

5. Ongoing development of the telemetry system

In the framework of an EU-ARCADIS project in Trans-Carpathia, 8 hydrological (HPT) and 5 hydrometeorological stations with satellite data transmission are planned for 2005. The DANCEE-DHI development programme is expected to produce 3 new hydrological (HPT) stations, i.e., 16 new stations are expected on top of the existing 20 (14+1+5). Thus, the total number of stations at the end of 2005 is expected to be 36 in Trans-Carpathia. Thirty-eight hydrological stations have been installed in the frame of the PHARE-CBC project (Flood prevention in the upstream Tisza river basin) and 5 new stations have been installed in the frame of the DESWAT (Destructive Water Abatement and Control of Water Disasters) project in the Romanian section of the Upper-Tisza basin. The uninterrupted operation of these stations has not been reached yet at the end of 2004. Five stations are equipped with water quality sensors. Measurements are taken in 5-minute intervals; data transmission is scheduled twice a day, but manual data retrieval can be initiated from the telemetry centre. Data collection is carried out by the Baia Mare telemetry centre and the data are transmitted to the Cluj Somes Tisa water dispatching centre.

Table 3. Evaluation of water level measurements at the hydrological telemetry stations

Stream	Station	Days with missing data (%) / score	Mean error / score	Maximum error / score	Days with error larger than 10 cm / score	RMSE / score	Total number of score
Tisza	Rakhiv	48.4/10	1/10	-57/5	14/7	4.1/3	35
Tisza	Tiachiv	2.0/40	5/5	142/0	154/0	10.6/0	45
Tisza	Khust	1.5/40	2/7	-52/5	147/0	10.9/0	52
Black-Tisza	Iasinia	46.3/10	-3/5	-66/5	12/7	4.8/3	30
White-Tisza	Luhi	67.0/0	1/10	-7/15	0/15	0.6/10	50
Tereshva	Usti Tshorna	23.2/30	-3/5	-45/5	95/0	5.4/3	43
Tereblia	Kolochovo	12.8/40	0/10	-31/7	18/7	3.8/5	69
Rika	Mizhgiria	3.5/40	-2/7	71/2	39/5	4.7/3	57
Borzhava	Dovge	4.0/40	1/10	-16/10	6/10	1.9/8	78
Uzh	Uzhgorod	21.2/30	-1/10	-14/10	21/5	2.7/5	60
Tisza	Tiszabecs	29.2/30	1/10	-19/10	15/7	3.7/5	62
Tisza	Tivadar	38.5/20	2/7	17/10	11/7	1.7/8	52
Tisza	Vás.namény	0.0/50	-2/7	-21/7	34/5	2.5/5	74
Tisza	Záhony	1.8/40	-3/5	20/10	61/0	2.8/5	60
Tisza	Dombrád	0.0/50	0/10	20/10	15/7	2.3/5	82
Tisza	Tiszabercel	0.0/50	1/10	-18/10	3/15	2.0/8	93
Túr	Garbolc	0.0/50	0/10	13/10	2/15	2.1/5	90
Túr	Sonkád	0.8/40	0/10	-15/10	3/15	2.0/8	83
Szamos	Csenger	0.0/50	-1/10	-41/5	38/5	3.9/5	75
Szamos	Tuny.matolcs	0.0/50	-2/7	20/10	2/15	1.5/8	90
Kraszna	Ágerdőmajor	0.0/50	0/10	-11/10	1/15	1.1/8	93
Kraszna	Kocsord	0.0/50	2/7	10/15	5/15	1.9/8	95
Túr-belvíz	Sonkád	0.8/40	-3/5	31/7	4/15	2.3/5	72
Lápi főcsat.	Nagyecsed	4.0/40	0/10	18/10	2/15	1.7/8	83
Belfő-csat.	Tiszabercel	0.0/50	1/10	-10/15	1/15	1.5/8	98
III. sz. ff.	Kántorjánosi	21.4/30	0/10	19/10	1/15	1.4/8	73
Lónyay-fcs.	Kótaj	0.0/50	0/10	11/10	2/15	1.5/8	93
Rétközi-tó	Sz.veresmart	10.6/40	3/5	20/10	41/3	1.0/10	68

6. Proposal for further development of the telemetry system

The future steps include the following points:

- Certain measures are necessary to maintain the reliable operation of the Hungarian – Ukrainian telemetry system. The number of missing data has to be reduced. There is a need for back up of the data transmission lines in Trans-Carpathia.
- Telemetry stations of agencies outside of the water management network will be integrated into the joint Hungarian – Ukrainian hydrological telemetry system.
- There is a need for additional groundwater, precipitation and air temperature stations, and the automated measurement of meteorological elements should be extended.
- There is an urgent need to harmonise Hungarian, Ukrainian and Slovak programmes of development with the corresponding Romanian programmes. Experts from these four countries concluded (NATO 2002) that there is a necessity to create a shared database, beside the national data bases, and that digital data exchange should also be maintained (Figure 4).
- Integration of the planned Slovak stations is foreseen in the POVAPSYS complex flood forecasting programme (40 telemetry stations) (NATO 2002).
- Archived data should be accessible, beside the current operational use, for other purposes with variable frequency.
- The telemetry system should concentrate on observations of such elements, which can be utilised by forecasting models.

References

Adler MJ, Amafteesei R, Corbus C, Ghioca M, Stancalie G (2002) The modernization of the system of measurement, storage, transmission and dissemination of hydrological data to various decision levels (ECMOSYM Project). Proceedings of the International Conference Preventing and Fighting Hydrological Disasters. November 21 and 22, pp 433-438

Bálint Z, Konecsny K (2002) Development of a Flood Alarm and Forecasting System on the Upper-Tisza River Basin. Proceedings of the 21st Conference of the Danube Countries on the hydrological forecasting and hydrological bases of water management. 2-6 September, Bucharest

Domahidy L, Puskás T (1998) National hydrological telemetring and warning system (in Hungarian) Vízügyi Közlemények/Hydraulic Engineering. Year LXXX, Number 1

Illés L (1995) Modernization of the remote sensing network at the Upper-Tisza Water Authority (in Hungarian) FETIVIZIG Nyíregyháza

Illés L (2002) Concept of the development of flood forecasting in the upper Tisza River (in Hungarian) Vízügyi Közlemények/Hydraulic Engineering. Year LXXXIV, Number 1

Illés L, Konecsny K (1995) Hydrological experiences of the Upper-Tisza flood in 1993 and an evaluation of efficiency of the flood forecasting system (in Hungarian). M.H.T. XIII. National Conference, Baja, 4-6 July

Illés L, Domokos L (1989) A Felső-Tisza vízrajzi távmérő rendszer működési tapasztalatai és fejlesztése. MHT. Vándorgyűlés II. füzet

Illés L, Konecsny K (2001) Flood forecasting (III). In: Upper Tisza Flood November 1998 (in Hungarian). FETIVIZIG-VIZITERV Nyíregyháza-Budapest, pp 77-126

Illés L, Konecsny K, Lucza Z (1999) Operation of the modernized remote sensing system in November 1998 (in Hungarian). MHT XVII. National Conference I, Miskolc 07-08 July, pp 55-67

Konecsny K (2003) Experiences gained with the development and operation of the network of automated river gauges along the Upstream-Tisza River (in Hungarian). Hidrol. Közl. /Journal of the Hungarian Hydrological Society 83(4):193-206

Konecsny K (2004) A felső-tiszai közös magyar-ukrán vízrajzi távmérő rendszer egy éves üzemelésének értékelése. Vízügyi Közlemények, LXXXVII. Évfolyam. 1-2. füzet

Pasche E, Kraßig S, Schlienger M (2002) Flood prevention strategies for the Upper Tisza river. Proceedings of the International conference preventing and fighting hydrological disasters. 21-22 November, Timisoara (Romania), pp 135-142

Puskás T, Karsai H (1974) Advances in the automation of the network serving hydrological forecasting and operation control in water management (in Hungarian). Hidrológiai Közlöny/Journal of the Hungarian Hydrological Society. No. 6

Stadiu F, Adler MJ (2002) Destructive Water abatement and control of water disaster (DESWAT Project). The International Conference Preventing and Fighting Hydrological Disasters. 21-22 November, Timisoara (Romania) pp 409-412

Szlávik L (1999) Tasks of preventing and fighting water hazards, concept of developing the national information system, implementation and operation (in Hungarian). Vízügyi Közlemények/Hydraulic Engineering. Year LXXXI, Number 1

Szlávik L, Buzás Z, Illés L, Tarnóy A (1997) International cooperation for managing the water resources of the Tisza River Valley (in Hungarian). Vízügyi Közlemények/ Hydraulic Engineering. Year LXXIX, Number 3

NATO (2002) Joint Ukraine-NATO project on flood preparedness and response in the Carpathian region. NATO, Brussels, Belgium

LAND USE MAP FROM ASTER IMAGES AND WATER MASK ON MODIS IMAGES

JUDIT KERÉNYI, MÁRIA PUTSAY[1]
Hungarian Meteorological Service, Budapest, Hungary H-1024

Abstract. In recent years, extreme floods occurred frequently in many parts of the world. One region, which suffers from flood damages on a regular basis, is the transboundary area of the White Körös and Black Körös rivers flowing from Romania into Hungary. For further improvement of flood management in this area, an international team was formed, with representatives of Hungary, Romania and USA and a project started on "Monitoring of Extreme Flood Events in Romania and Hungary Using Earth Observation (EO) Data" in the framework of the NATO Science for Peace (SfP) Programme. Remote sensing technologies, GIS (Geographic Information System) database merging the surface and remote sensing data, numerical hydrological predictions and the traditional means can greatly improve the management of flood hazards. The paper provides a brief overview of the satellite data processing tools needed for the project, developed at the Hungarian Meteorological Service (HMS). Our task was to produce a high-resolution land use map for the study area and develop a method for floodwater detection. The land use map will be used as background information with topographical, physiographical and hydrological parameters in the GIS database. We have derived it by applying 15-m spatial resolution ASTER images. The method for the land cover / land use mapping included two steps: (a) Geo-referencing and (b) Data classification. During post-processing, the classes were merged into 8 groups: 4 agricultural areas, and forest, water, cloud, and shadow. Urban areas could not be separated in a discrete class so they had to be classified manually. Two methods were developed for water detection using

---
[1] To whom correspondence should be addressed. Maria Putsay, Hungarian Meteorological Service, H-1024 Budapest Kitaibel Pal ul, Hungary; e-mail: putsay.m@met.hu

multispectral MODIS images. Both methods are threshold techniques. One is automatic, quick, but less accurate, because it may contain misdetected cloud shadow pixels. The other algorithm is a quasi-automatic method, more accurate but interactive correction is needed if the investigated area is partly covered by clouds. Water masks were produced for the period of the spring 2000 flood. The water vectors concerning successive days were visualised in a common map (and on the land use map as well) to see how the inundated area varied.

Keywords: flood management; remote sensing; land use map; detecting flood water; ASTER; MODIS

1. Introduction

In recent years large and extreme floods occurred frequently in many parts of the world. One region, which suffers from flood damages on a regular basis, is the transboundary area of the White Körös and Black Körös rivers flowing from Romania into Hungary. Floods in this area typically start in the mountainous terrain of the upper parts of the basin in Romania and propagate to the plains in Hungary. Recent floods in this area include the two spring 2000 floods, which caused damages on the Romanian territory. On the Hungarian territory, a particularly notable flood occurred in the summer of 1980.

There exists a close co-operation between Romania and Hungary concerning the flood protection. The information provided by Romania to Hungary (downstream) is presently based entirely upon the ground (in-situ) data, mainly collected by the non-automatic hydrometeorological stations. Such data are somewhat limited in terms of spatial distribution, temporal resolution, and the speed of collection and transmission.

Recognising the need for further improvement of flood management in this area, an international team was formed, with representatives of Hungary, Romania and USA, and a project started on "Monitoring of Extreme Flood Events in Romania and Hungary Using Earth Observation (EO) Data" in the framework of NATO Science for Peace (SfP) Programme.

The success of risk management largely depends on the availability, dissemination and effective use of timely information. In flood risk management remote sensing technologies and the numerical hydrological predictions are used together with the traditional means, and can greatly improve the management of flood hazards. The SfP project aims to provide such an efficient and powerful flood monitoring tool. A GIS (Geographic Information System) database is organised merging the surface and remote sensing data.

The paper that follows provides a brief overview of the satellite data processing tools needed for the project. This work was done at the Satellite Research Laboratory of the Numerical Weather Prediction Division of the Hungarian Meteorological Service (HMS). Our task was to produce a high-resolution land use map of the study area and develop a method for floodwater detection.

2. Study area: Körös basin

The study area is the Körös transboundary basin located across the Romanian–Hungarian border, with a total area of 26,600 km² (see Fig. 1).

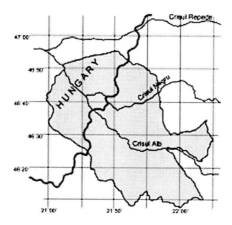

Fig. 1. The study area, the Körös basin

In Romania, 38% of the catchment area is mountainous, 20% comprises hills and 42% plains. About 30% of the Romanian part is forest. On the Hungarian side, the catchment relief represents plains. Annual precipitation ranges from 600-800 mm/year in the plain and plateau areas to over 1200 mm/year in the mountainous areas of Romania. This precipitation distribution can be explained by the fact that humid air masses brought by fronts from the Icelandic Low frequently enter this area. The orography of the area (Apuseni Mountains) amplifies the precipitation on the western side of the mountain range. The Körös Rivers Basin frequently experiences large precipitation amounts over short durations and the frequency of such events seems to be increasing in recent years.

The Romanian part of the basin is of greater interest with respect to flood formation, because of the high rates of mountain runoff.

3. Development of the GIS database

The database contains a spatial geo-referenced information ensemble, with satellite images, thematic maps, time series of meteorological and hydrological parameters, and other data. The GIS database is connected with the hydrological database.

The GIS database contains the following info-layers: sub-basin and basin limits, land topography, the hydrographic network of dykes and canals, transportation network (roads, railways), municipalities, meteorological station network, raingauge network, hydrometric station network, land cover/land use (updated from satellite images), and water mask of the flooded area derived from the previous flood periods.

4. Image processing for flood analysis

HMS is dealing with producing the land cover/land use map and detecting floodwater. In this work medium spatial resolution TERRA/MODIS and high-resolution ASTER images were applied.

4.1. LAND USE MAP DERIVED FROM ASTER IMAGES

To get information about the land cover for the investigated area we derived a land use map from ASTER images. This information will be used as background information with topographical, physiographical and hydrological parameters in the GIS database.

ASTER is a sensor onboard the Terra (EOS AM) and Aqua (EOS PM) satellites. The ASTER acquires 14 spectral bands. Aster images were found suitable for producing detailed maps of land cover/land use. The 15-m spatial resolution of visible and near infrared bands was used: band1: 0.56 μm, band 2: 0.66 μm and band 3N: 0.8 μm.

For the land classification we used the ENVI™ software. The land use map was derived from images taken in 2002 and 2003.

The method for the land cover/land use mapping included two steps: (a) Geo-referencing and (b) Data classification. For the classification both supervised and unsupervised methods were tested. Better results were obtained with the unsupervised classification, in which different numbers of classes and iterations were tested (90 classes and 40 iterations were chosen). During the post-processing the classes were merged into 8 groups: 4 agricultural areas, forest, water, cloud, and shadow. The urban areas could not be separated into a discrete class so they had to be classified manually.

The classification was performed for four ASTER images to cover the Hungarian part of the investigated area and a mosaic land use map was made from them.

The raster type images will be vectorised and evaluated for the GIS data base. The Romanian part of the map made by colleagues at the National Meteorological Administration and was merged with our map. The land use maps will serve to calculate flood damages.

4.2. WATER DETECTION USING MODIS IMAGES

MODIS (Moderate Resolution Imaging Spectroradiometer) is another sensor onboard the Terra (EOS AM) and Aqua (EOS PM) satellites. Terra MODIS and Aqua MODIS view the entire Earth surface every 1 to 2 days. MODIS is a passive imaging spectroradiometer, acquiring data in 36 spectral bands.

The data available from MODIS are highly suitable for flood warning and management, because they are available in real time (or nearly real time), can be rapidly processed and disseminated, cover a wide area, and are abundant and inexpensive, which is important when dealing with floods of longer durations. MODIS provides very accurate geo-coding, clouds can be removed by composition techniques, and the data are distributed via the EOS Gateway.

For our purpose we used Level 1B data, visible bands 1 (0.66 µm) and 2 (0.87 µm) with 250 m spatial resolution and near infrared channel data (1.6 µm) with 500 m spatial resolution.

MODIS Level 1B data from the Terra Satellite were downloaded from the EOS Data Gateway, Land Processes Distributed Active Archive Center (LP DAAC) World-Wide Web site: http://edcimswww.cr.usgs.gov/pub/imswelcome/.

The classical method of detecting floodwater in satellite images is the following: visualise the image (preferably a composite image - visualising more channel data simultaneously), highlight it and delineate the inundated area manually. This method is work- and time-consuming. Our aim was to develop an automatic method to create water mask in a faster, easier and more objective way.

Typically, the NDVI (Normalised Difference Vegetation Index) and/or the 0.87 µm reflectance are used for water detection as these have low values for clear water (and these data have the highest spatial resolution among the MODIS data). The 0.87 µm reflectance and NDVI is much lower for water surfaces than for vegetation or bare soil, so this is a good tool for detecting pixels totally covered by clear water. But unfortunately for mixed pixels partly covered by vegetation the NDVI and the 0.87 µm reflectance value can strongly increase, and they are sensitive to vegetation or turbid water. NDVI was expressed as

$$NDVI = (R0.87 - R0.66)) / (R0.87 + R0.66)$$

where R0.66 and R0.87 are the reflectance values in channels 0.66 µm and 0.87 µm.

Using only one test for NDVI is not enough, and one has to apply more tests to detect not only the pixels formed 100% by clear water, remote from snow, clouds and cloud shadows. The aim was to find an automatic algorithm to detect pixels that were either 100% water or partially water covered.

At HMS the staff of the Satellite Research Laboratory already had experience with automatic cloud detection methods for NOAA images (Kerényi et al. 1995). This algorithm has been routinely used at the Hungarian Meteorological Service for years. As an analogue for this cloud detection method we started to develop a method for water detection using multispectral MODIS images.

The threshold technique works for each pixel separately. It investigates the spectral characteristics of the pixels individually, without taking into account the adjacent pixels. The method consists of more tests, each test examines whether the measured value in a channel (or a value calculated from more channel data) is greater (or smaller) than a predefined threshold. If a given combination of the tests fulfils the detection objective, then we detect the pixel as covered by water.

The main thing is to find the appropriate tests (with physical meaning) and the appropriate thresholds. The thresholds should be general in the sense that they should not vary from image to image (some seasonal variation can occur, but this should also be predetermined).

To improve chances of finding general thresholds the most important thing is to use calibrated rather than raw data. Water detection algorithm works with channels measuring reflected solar radiation, so it is very important to eliminate their variation from the illumination conditions (the solar elevation has seasonal variation at the satellite overpass). This elimination can be done by using calibrated values, namely solar zenith angle corrected reflectances.

Next task was to find the tests, and their accurate combination. As we have already stated NDVI and the 0.87 µm reflectances are useful for clear water detection. The channel 1.6 µm is also very useful for water detection, because it is characterised by the lowest water reflectance, mostly independent of turbidity, which is typical for floodwaters. The reflectance in 1.6 µm channel is less affected by vegetation, than the 0.87 µm channel reflectance. Oancea and colleagues made reflectance measurements at several wavelengths and they confirmed this behaviour (Oancea et al. 2003). The spatial resolution in the 1.6 μm channel is only half as good as for the 0.87 μm channel, but it is so informative that we decided to use it as the main test (for water the 1.6 μm reflectance should be less than a threshold). But this 1.6 μm channel test cannot be used alone either, as the 1.6 μm channel allows separating water from lowland areas, but not from snow in mountainous areas (snow is bright in 0.87 μm channel, but dark in 1.6 μm channel). To mask out snowy areas, a threshold on the 0.87 μm channel was used, and the 0.87 μm reflectance should be less than this threshold. For water both tests should be fulfilled at the same time.

The other problems encountered were caused by orographic and cloud shadows, and by melting snow. Their spectral characteristics were quite similar to those of water, so only partial separation was achieved. Large, clear water surfaces without vegetation have NDVIs less than -0.2 and can be easily separated from shadows. However, using this threshold would mask out not only the shadows but also many pixels containing water, such as the pixels with high vegetation fractions, turbid waters, and mixed pixels along the coasts. To avoid the loss of such pixels, an NDVI threshold greater than -0.2 had to be used (we made many tests to find the best NDVI threshold; it is seasonal dependent). We decided to use three tests: the first for ch 1.6, the second for ch 0.87 and the third for NDVI; for water the conditions imposed by all the tests should be fulfilled.

In 2003 we established the basis of the method, chose the channels to use in the tests, started to look for the appropriate equations and thresholds, and tested the new method on four MODIS images (Putsay 2003).

In 2004 we continued to test our automatic water detection method on a number (11) of images to find the most powerful method and the best thresholds (Putsay 2004). Good thresholds must be tested on many images reflecting many circumstances, in different seasons. We also investigated which composite images are better to use for validating the method.

We had to make compromises. We could not separate completely the water and cloud shadow pixels automatically. In a few cases cloud shadows were misdetected as water pixels. Thus, we decided to develop the water detection method in two ways:

- First is an automatic method, which is quick, but it can misdetect some shadow pixels as water and it does not detect some water pixels. We worked on it in 2004.

- The other method is a quasi-automatic method, which is more accurate, but it needs more time, and interactive correction. We were working on this method in 2005.

In 2004 our efforts focused mainly on finding the optimal thresholds for the separation of water from shadows. These thresholds can be used if we do not have time for a posterior manual correction. We have found the following thresholds:

$$(R1.6 < 1\%) \text{ or } [(R1.6 < 5\%) \text{ and } (R0.87 < 13\%) \text{ and } (NDVI<T)]$$

where T varies between −0.2 and +0.2 depending on the season and the cloudiness, and R1.6 and R0.87 are the reflectance values in channels 1.6 μm and 0.87 μm, respectively. Here we introduced a modification of the above equation as well. If the 1.6 μm reflectance is very low (less than 1%), no additional tests are needed.

In 2005 we worked on the quasi-automatic method, which is more accurate, but it needs interactive correction (removal of misdetected cloud shadow vectors manually). We were working with MODIS images taken during the spring 2000 flood on the Körös Rivers. We worked with images properly corrected for solar zenith, bow-tie and geo-correction.

We have developed a quasi-automatic method by modifying the thresholds and the governing equation. The aim is to get accurate water mapping. After the automatic part we have more shadows but an accurate water mask. First we develop the water mask raster image, then we convert it into vectors, and afterward we remove the misdetected cloud shadow vectors manually. The new equation and the thresholds are as follows:

$$(R1.6 < 1\%) \text{ or}$$

$$[(R1.6 < 10\%) \text{ and } (R0.87 < 18\%) \text{ and } (NDVI < 0.4)] \text{ or}$$

$$[(R1.6 < 15\%) \text{ and } (R0.87 < 15\%) \text{ and } (NDVI < 0.0)].$$

For a lower NDVI value we permit higher R1.6 values to keep pixels with less water portion.

In figures 2 and 3 water masks derived for four successive flood days are shown. We visualise two water vectors corresponding to different days together to show the variation of the inundated area. Water vectors were overlaid on a channel 0.87 μm image as a background (which was taken on the 21st of April and the 2nd of May 2000, respectively).

4.3. WATER MASKS ON THE LAND USE MAP

In Figure 4 the land use map (derived from ASTER image) is seen as background and water masks corresponding to two days are overlaid. The gray water mask corresponds to the 21st of April and the black one to the 2nd of May 2000. Such investigation provides information about the inundated area and helps estimate the flood damage.

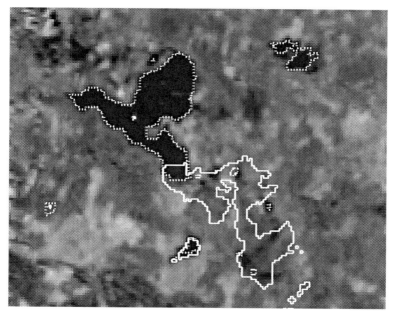

Fig. 2. Water masks derived for two days using MODIS images, visualised on a channel 0.87 μm image (taken on the 21st of April 2000). The solid line water mask corresponds to the 9th, the dotted line water mask to the 21st of April

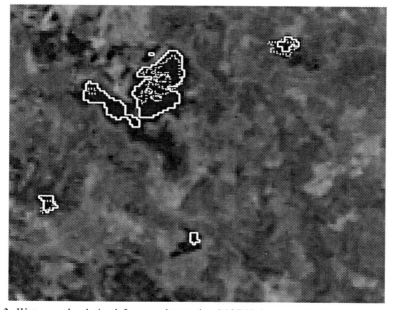

Fig. 3. Water masks derived for two days using MODIS images, visualised on a channel 0.87 μm image (taken on the 2nd of May 2000). The solid line water mask corresponds to the 2nd, the dotted line water mask to the 11th of May

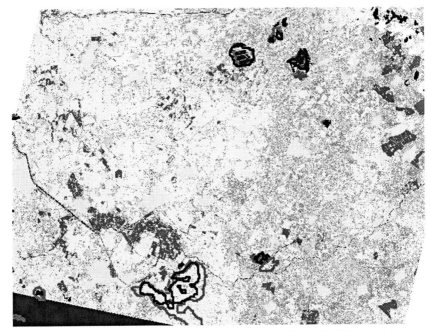

Fig. 4. Water masks overlaid on the land use map (derived from ASTER image). The gray water mask corresponds to the 21st of April and the black one to the 2nd of May 2000

5. Conclusions and future plans

We have reviewed the needs for the satellite data processing tools and the work done at the Hungarian Meteorological Service. Our task was to produce a high-resolution land use map of the study area and to develop a method for floodwater detection. The land use map will be used as background information with topographical, hypsographical and hydrological parameters in the GIS database. The map was derived from 15-m spatial resolution ASTER images in two steps: (a) Geo-referencing and (b) Data classification. During the post-processing the classes were merged into 8 groups: 4 agricultural areas, forest, water, cloud, and shadow. The urban areas could not be separated in a discrete class so they had to be classified manually.

Two threshold-technique methods were developed for water detection using multispectral MODIS images: one is automatic, quick, but less accurate; it may contain misdetected cloud shadow pixels. The other algorithm is a quasi-automatic method, more accurate but an interactive correction is needed if the investigated area is partly covered by clouds. Water masks were produced for the spring 2000 flood event. The water vectors concerning successive days were visualised in a common map (and on the land use map as well) to show how the inundated area varied.

Further plans (2005) include producing the water vector layers for all 11 MODIS images corresponding to the spring 2000 flood event.

Acknowledgement

This work is supported by the NATO Science for Peace (SfP) Programme project on "Monitoring of Extreme Flood Events in Romania and Hungary Using Earth Observation (EO) Data". We are grateful to the staff of the Dartmouth Flood Observatory (Hanover, USA), G. Robert Brakenridge, Emily S. Bryant and Elaine Anderson, who helped our work by their advice and provision of data. The authors are also indebted to the staff of the Remote Sensing and GIS Laboratory (NMA, Romania) for their help, and above all to Gheorghe Stancalie for his organisational work.

References

Kerényi J, Szenyán IG, Putsay M, Wantuch F (1995) Cloud detection on a threshold technique for NOAA/AVHRR images for the Carpathian Basin. Proceedings of the 1995 Meteorological Satellite Data Users' conference, 4-8 Sept, Winchester, United Kingdom, pp 565-569

MODIS World-Wide Web site: http://edcimswww.cr.usgs.gov/pub/imswelcome/

Oancea S, Alecu C, Bryant E (2003) MODIS water classification report, (unpublished report) prepared for/at the Dartmouth Flood Observatory

Putsay M (2003) Creating a water mask using a threshold technique on multispectral MODIS images. Report of visit of Maria Putsay to Dartmouth Flood Observatory, November 25

Putsay M (2004) Creating water mask using a threshold technique on multispectral MODIS images (http://nato.inmh.ro/MEDIA/nato2_presentations.html)

CHAPTER 2   FLOOD FORECASTING AND MODELLING

APPLICATION OF METEOROLOGICAL ENSEMBLES FOR DANUBE FLOOD FORECASTING AND WARNING

GÁBOR BÁLINT[1], ANDRÁS CSÍK, PÉTER BARTHA, BALÁZS GAUZER
Water Resources Research Centre, VITUKI, Kvassay J. út 1., Budapest, H-1095, HUNGARY

IMRE BONTA
Hungarian Meteorological Service, OMSZ, Kitaibel Pál u.1., Budapest, H-1024, HUNGARY

Abstract. Flood forecasting schemes may have the most diverse structure depending on the catchment size, response or concentration time, and the availability of real-time input data. The centre of weight of the hydrological forecasting system is often shifted from hydrological tools to the meteorological observation and forecasting systems. At lowland river sections simple flood routing techniques prevail where accuracy of discharge estimation might depend mostly on the accuracy of upstream discharge estimation. In large river basin systems both elements are present. Attempts are made to enable the use of ensemble of short and medium term meteorological forecast results for real-time flood forecasting by coupling meteorological and hydrological modelling tools.

Keywords: real time flood forecast, hydrological and meteorological ensembles, the Danube River, quantitative precipitation forecast, gridded fields, semi-distributed models

---

[1] To whom all correspondence should be addressed. Gabor Balint, Water Resources Research center, VITUKI, Kvassay J. ut 1., Budapest, H-1095, Hungary; e-mail: balint@vituki.hu

1. Components of the flood forecasting system

Flood forecasting provides essential information for flood defence. This type of information is handled within the national Flood Management Information System in Hungary. Transit flow originating from the upstream parts of the upper and central Danube Basin dominates hydrological regime of Hungarian rivers, and consequently hydrological systems cover an area of more than 300,000 km^2 mostly outside of the national boundary. The central unit of the forecasting system is operated for the 210,000 km^2 catchment of the Danube River upstream of the southern border, delineated by the cross-section near the town of Mohács. Separate units deal with tributaries the Tisza and the Dráva. All three units are managed by the National Hydrological Forecasting Service (NHFS) within the VITUKI Environmental Protection and Water Management Research Institute.

The present study concerns only the Danube proper, which is the receiving water body of German, Austrian and Slovak tributaries. The Danube forecasting system is linked to the METINFO system of the Hungarian Meteorological Service providing meteorological forecasts and observations. The hydrological data collection and pre-processing system linked to similar services of the Danube countries handles part of the meteorological data and water level and discharge data of more than 90 hydrological observation sites. The NHFS modelling system performs data assimilation and produces 12-hour water level and discharge forecast 6 days ahead for 46 forecast stations.

1.1. METINFO - METEOROLOGICAL FORECASTS AND OTHER PRODUCTS

The European Centre for Medium-Range Weather Forecasts (ECMWF) products are used up to 10 days ahead. The ECMWF is an international organisation supported by 25 European States, based in Reading, west of London, in the United Kingdom. Products from the deterministic atmospheric model used mostly for hydrological purposes are: fields of 10-m U- velocity (10U); 10-m V-velocity (10V); 2-m temperature (2T); 2-m dew point (2D); 2-m max temperature (MX2T); 2-m min temperature (MN2T), Total cloud cover (TC), and Total precipitation (TP) in GRIB code form (40 x 40 km grid). Further Atmospheric Products from the Ensemble Prediction System are also utilised, namely 2-m temperature (2T), 2-m max temperature (MXT), 2-m min temperature (MNT), and Total precipitation (TP) of the basic ensemble forecasts in 80 x 80 km grid.

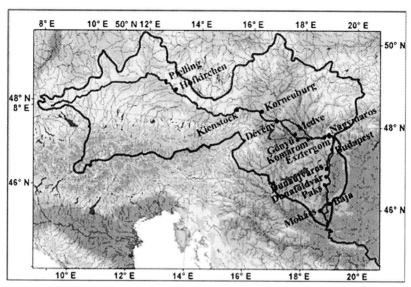

Fig. 1. Upper and Central Danube forecast stations

The upper and central part of the Danube basin is covered by the forecasts LACE (Limited Area Modelling for Central Europe) model. LACE is the name of the co-operative venture among Central-European meteorological and hydro-meteorological Institutes (Austria, Croatia, Czech Republic, Hungary, Slovakia, and Slovenia), which aims a common development and operational exploitation of a limited area numerical weather prediction model. The model used is the ALADIN spectral limited area model developed in an international collaboration (with the participation of 14 countries) led by Meteo France. Recently the operational version of the ALADIN model called ALADIN/LACE was exploited in Prague and Budapest for a domain covering Continental Europe. The model is launched twice a day (at 00 and 12 UTC) giving 48-hour forecasts. The products of the model are transferred through telecommunication lines to the disposal of the Member States. The received products are intensively used for short-range forecasts issued by the Member States. Some of the Members run the local version of the ALADIN model centred for their territory providing an even more precise and exact short-range forecasts for the forecasters. The Hungarian Meteorological Service (HMS) runs such a local model utilising the output of the central model as boundary conditions. The central model has the spatial resolution of 12 x 12 km grid, while the local ones use 6.5 x 6.5 km resolution. Time-step of model output is 400 s, i.e. 432 time steps within the 48 hours maximum lead time. The Budapest window covers the whole Carpathian Basin including the Tisza catchment.

1.2. THE HYDROLOGICAL MODELLING SYSTEM

The NHFS GAPI/TAPI modelling system has been developed at NHFS VITUKI. The conceptual, partly physically-based GAPI model serves for simulations and forecasting of flow for medium and large drainage basins. The lumped system consists of sub-basins and flood routing sections. In the course of a decade of development and upgrading the forecasting package has grown into a complex tool containing snow accumulation and snowmelt, soil frost, effective rainfall, runoff and flood routing modules, extended with statistical error correction - continuous updating, and hydraulic 'empirical backwater effect' modules.

The Discrete Linear Cascade Model (DLCM) developed by Szöllösi-Nagy (1982) utilising an approach similar to the one reported by Szolgay (1984) serves for the routing of flow components and channel routing (Figure 3). First version of the complex GAPI model with modular structure was designed by Bartha et al. (1983). The choice of the model was proved by a number of inter-comparison studies (WMO 1992). The first model version was extended for a snowmelt module (Gauzer 1990) and the complexity of the system was gradually increased. The backwater module utilises simplifications similar to those suggested by Todini and Bossi (1986).

The large number of nodes (69) makes the system in fact semi-distributed in basin scale. Out of the total number of nodes, 46 are related to forecast stations. Real-time mode runs are carried out in 12 hourly time steps. Input/output values and state variables of the precipitation - runoff modules are integrated over sub-basins as a weighted or simple arithmetic average of station or grid values. A special procedure was designed to interpolate sparse grid data of ECMWF products. The rapid growth of resolution and forecast range of numerical weather prediction models enabled the use of their results for hydrological forecasting purposes. Unfortunately Quantitative Precipitation Forecasts remain the most uncertain elements of any meteorological prediction. The growth of computational power available for meteorological centres allowed the introduction and real time application of ensemble techniques capable of dealing with such uncertainty (Table 1).

Table 1. Improvement of Numerical Weather Prediction Models

Improvement of Numerical Weather Prediction models		
Year	ECMWF / Global Models	Regional Models / LAMs
1985	200 km	50 Km
2003	40 km (80km)	6 km
200?	~20 km	~ 1 km

2. The use of meteorological ensembles for flood forecasting

2.1. TESTING ON AN EXTREME EVENT

Within the 5th Framework Programme of the European Commission, a shared cost action called European Flood Forecasting System (EFFS) has been funded. The goal of the project was the setup and semi-operational testing of a continental-scale flood forecasting system for major river basins in Europe. The principal research aim was to explore the possibility to extend the lead-time of the flood warning process up to 10 days into the future. This was to be achieved with the use of numerical weather forecasts. Various deterministic and ensemble forecasts delivered by national and international meteorological services have been used within the system to drive a sequence of hydrological rainfall-runoff models and hydraulic models for principal river systems or selected pilot basins. The weather forecasts were downscaled through nesting form global circulation models to high resolution local models (Gouweleeuw et al. 2004). The raster based hydrological model LISFLOOD (De Roo et al. 2000) and other hydrological models, such as HBV (Bergström 1976), TOPKAPI (Ciarapica and Todini 2002), the FloRIJN model system consisting of the HBV model and the hydrodynamic SOBEK model, the Hron model (Szolgay et al. 2003), the NAM model, the VIDRA model and the GAPI-TAPI system have also been implemented and tested for the particular basins.

Operational use of the above system often revealed the uncertainty of QPF taken into consideration while calculating expected Danube hydrographs. To test the feasibility of the use of meteorological ensembles a forecasting experiment was designed. The aim of the investigation was also to assess how much the prior estimates of uncertainty are controlled by the selected approach.

ECMWF archived forecast arrays were retrieved using partly the standard data ingestion part of the NHFS system. Fifty-two sets of input data arrays were produced using the 50 ECMWF ensemble elements; additionally deterministic and control runs were also carried out. Control run input is identically produced as the deterministic one, but instead of the 40 km resolution, 80 km resolution is used. The period of July-August 2002 was selected for this study and included the period of the August flood.

The two flood waves were induced by torrential rains on Aug. 7 and 8, and 11-13, respectively. The precipitation of the first rainfall caused high area average especially over the upper reach of the Danube. The peak of the first flood wave reached Budapest on Aug. 11, with a moderate flow rate. The following second rainfall period resulted in extreme floods on the Austrian tributary the Salzach, and on smaller streams in the Upper Austria region with destructive consequences. The peak reached Vienna on the 15th of August. Historical high flows occurred along the Slovak-Hungarian section of the Danube.

The actual hindcasting took place after the cycles of input data preparation were completed. Due to the limitations of the NHFS system, only 1-6 day ahead forecast was calculated for the period from the 15th of July to the 31st of August. Forecast was calculated at daily frequency using 12-hour time steps. That followed the running of the forecast-system. The calculation of the ensemble forecast is time demanding; that is why the standard operational procedure was modified, flags and options were reduced to produce the batch type of processing (altogether, more than 5,000 runs were performed, which is equivalent to more than 6 years of 'real time' activity). Twelve-hour time step 1-6 day ahead forecasts were calculated for 46 forecast stations. Out of those 21 were analysed comparing forecast results to observed hydrographs.

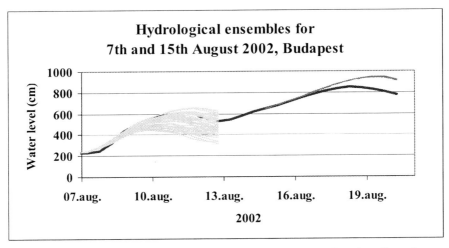

Fig. 2. Hydrological ensembles indicating great dispersion on the 7th and no dispersion on the 15th August 2002, at the Budapest gauge

2.2. RESULTS

Different behaviour of the system can be observed in Figure 4. The 'torch' diagrams indicate the differences between situations with uncertain and univocal QPF ensembles. The 50-element ensemble of hydrological forecast reflects the great uncertainty associated with the August 7th forecast. The other case shows almost no differences on the 15th August. The QPF error related forecast error is negligible in this case and the differences between observed water level/discharge values are explained by the relatively poor performance of the flood routing part of the system. Unfortunately for the hydrological users the cases of QPF with high certainty are usually associated with no rain situations.

Figures 3a and 3b show main features of different sets of hydrological ensembles for gauging stations Devin and Budapest. The specific Box-Whisker diagrams indicate beside observed hydrographs forecast arrays showing minimum and maximum values of 50 element ensembles while quartiles above and below the mean values are indicated by wider boxes. Forecast is indicated for 24, 72, and 144 hours lead times. Even this two gauge comparison is sufficient to indicate the impact of growing travel time along the 200 km reach which is expressed in higher accuracy of forecast at the lower (Budapest) section while the rainfall - runoff module has higher weight at Devin, and consequently forecast error is higher at the upstream station. The natural increase of error with the lead time can also be followed. Upstream rainfall induced flood waves impact on the Budapest section only 2-3 days after the rainfall (or snowmelt) event occurs. The limit of predictability is reached at Devin on the 6th day; however, ensemble means still give some useful information.

Forecast types are compared with each other in Tables 2 and 3. The basic comparison was performed by applying the so-called efficiency coefficient, a skill score widely used in hydrology. This table refers to the period from July 21 to Aug. 31, 2002 and the colours (shades) show the forecast efficiency. In the case of the upper section and higher lead time the mean of 50 members is better because the precipitation forecast is dominant in the hydrological forecast with a long (3-6 days) lead time. Down to the stream – when the flow routing is dominant – the operative forecast gives the higher skill score values.

Figure 4 shows distribution of 1-6 day forecasts compared to the climatic distribution for Aug. 14, 2002 at the Budapest gauge. For decreasing lead time, distribution functions are more and more narrow reflecting the gradually decreasing uncertainty.

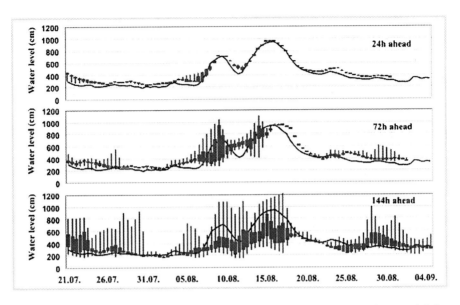

Fig. 3a. Hydrological ensemble sequences for the Devin gauge; quartiles of 1, 3 and 6 day forecasts

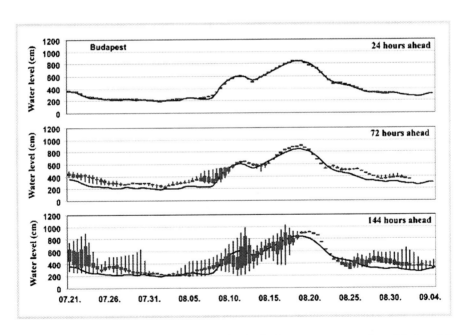

Fig. 3b. Hydrological ensemble sequences for the Budapest gauge; quartiles of 1, 3 and 6 day forecasts

Table 2. Efficiency index for the period 21st July – 31st August 2002

Station / Lead time	24 hours	48 hours	72 hours	96 hours	120 hours	144 hours
Pfelling	0.77	0.82	0.79	0.74	0.69	0.64
Hofkirchen	0.86	0.85	0.83	0.80	0.73	0.66
Kienstock	0.85	0.77	0.87	0.85	0.78	0.73
Korneuburg	0.79	0.80	0.86	0.84	0.77	0.71
Devin	0.95	0.88	0.88	0.89	0.85	0.78
Medvedovo	0.93	0.89	0.82	0.87	0.86	0.82
Gönyü	0.95	0.93	0.87	0.88	0.88	0.85
Komárom	0.95	0.92	0.88	0.89	0.90	0.87
Esztergom	0.97	0.96	0.92	0.86	0.89	0.88
Nagymaros	0.98	0.98	0.96	0.90	0.89	0.89
Budapest	0.97	0.97	0.96	0.92	0.88	0.89
Dunaújváros	0.98	0.97	0.97	0.94	0.89	0.89
Dunaföldvár	0.98	0.96	0.96	0.95	0.91	0.87
Paks	0.99	0.98	0.96	0.95	0.92	0.86
Baja	0.98	0.98	0.97	0.96	0.93	0.88
Mohács	0.98	0.98	0.97	0.97	0.95	0.90

Legend: Mean of 50 members / Operative

Table 3. Nash and Sutcliffe criterion for the period 21st July – 31st August 2002

Station / Lead time	24 hours	48 hours	72 hours	96 hours	120 hours	144 hours
Pfelling	0.91	0.79	0.62	0.40	0.21	0.04
Hofkirchen	0.95	0.86	0.72	0.57	0.37	0.18
Kienstock	0.93	0.70	0.72	0.66	0.54	0.37
Korneuburg	0.92	0.77	0.76	0.70	0.55	0.37
Devin	0.98	0.89	0.81	0.78	0.69	0.53
Medvedovo	0.98	0.92	0.78	0.78	0.74	0.62
Gönyü	0.99	0.95	0.86	0.82	0.79	0.67
Komárom	0.99	0.95	0.88	0.84	0.82	0.72
Esztergom	1.00	0.98	0.93	0.81	0.81	0.75
Nagymaros	1.00	0.99	0.96	0.87	0.80	0.76
Budapest	1.00	0.99	0.96	0.90	0.79	0.76
Dunaújváros	1.00	0.99	0.97	0.93	0.82	0.78
Dunaföldvár	1.00	0.99	0.97	0.94	0.86	0.75
Paks	1.00	0.99	0.98	0.95	0.88	0.75
Baja	1.00	0.99	0.98	0.96	0.91	0.81
Mohács	1.00	0.99	0.98	0.98	0.94	0.86

Legend: Mean of 50 members / Operative

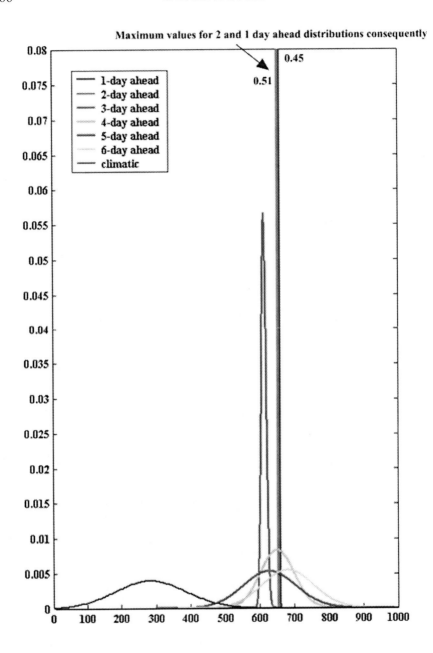

Fig. 4. Comparison of distributions, climatic and 1-6-day water level forecasts, for Aug. 14, 2002, in Budapest

3. Conclusions

The forecasting experiment proved that the use of meteorological ensembles to produce sets of hydrological predictions increased the capability to issue flood warnings. The NHFS system can be used for such a purpose; however, for real-time use the linkage between meteorological and hydrological modules should be considerably reviewed. The set of more than 5,000 model runs for the August 2002 extreme flood event was executed within a reasonable time period. Further important findings are:

- Appropriate decision support rules are needed to utilise the array of flood forecasts for flood management and warning purposes;

- The proper estimation of the contribution to forecast error by different modules of the system may help better understand expected uncertainty of the forecast;

- Any future forecasting exercise should include longer period of low flow or medium flow period to have proper estimates of 'false warning' types of errors.

Acknowledgment

This work has been partly supported by the Hungarian Research and Development Project: "Flood Risk Analysis", NKFP 3/067/2001. Authors also acknowledge gratefully the support of the European Commission provided for the European Flood Forecasting System Project EVG1-CT-1999-00011.

References

Bartha P, Szőllősi-Nagy A, Harkányi K (1983) Hidrológiai adatgyűjtő és előrejelző rendszer. A Duna. Vízügyi Közlemények 65(3):373-388

Bergström S (1976) Development and application of a conceptual runoff model for Scandinavian catchments, Swedish Meteorological and Hydrological Institute, Report RHO, No. 7, Norrköping, Sweden

Ciarapica L, Todini E (2002) TOPKAPI: A model for the representation of the rainfall-runoff process at different scales. Hydrological Processes **16(2)**: 207-229

Collier C, Krzysztofowicz R (2000) Quantitative precipitation forecasting. Journal of Hydrology 239 (1-4):1-2

De Roo A. (and 20 others) (2003) Development of a European flood forecasting system. International Journal of River Basin Management 1(1):49-59

De Roo A, Wesseling CG, Van Deursen WPA (2000) Physically based river basin modeling within a GIS: The LISFLOOD model. Hydrological Processes 14:1981-1992

Gauzer B (1990) A hóolvadás folyamatának modellezése (Modelling of the processes of snowmelt). Vízügyi Közlemények 72(3):273-289

Gouweleeuw B, Reggiani P, De Roo A (2004) A European flood forecasting system EFFS. Final Report, JRC, WL Delft Hydraulics, Italy

Szolgay J (1984) Modellierung von Flubstrecken mit seitlichen Zuflüssen mit der linearen Speicherkaskade, XII. Konf. Donau., Bratislava, pp 381-390

Szolgay J, Hlavčová K, Kohnová S, Kubeš R, Zvolenský M (2003) Influence of antecedent basin saturation on extreme floods in the upper Hron River Basin. Acta Hydrologica Slovaca 4(3):183-189

Szöllősi-Nagy A (1982) The discretization of the continuous linear cascade by means of state space analysis. Journal of Hydrology 58(3-4):223-236

Todini E, Bossi A (1986) PAB (Parabolic and Backwater) an unconditionally stable flood routing scheme particularly suited for real time forecasting and control. Journal of Hydraulic Research 24(5):405-424

WMO (1992) Simulated real-time intercomparison of hydrological models, Operational Hydrological Report No. 38, World Meteorological Organization, WMO No. 779, Secretariat of WMO, Geneva, Switzerland

COUPLING THE HYDROLOGIC MODEL CONSUL AND THE METEOROLOGICAL MODEL HRM IN THE CRISUL ALB AND CRISUL NEGRU RIVER BASINS

RODICA MIC[1], C. CORBUS
National Institute of Hydrology and Water Management
Bucharest, Romania

V.I. PESCARU, LILIANA VELEA
Romanian Meteorological Administration, Bucharest,
Romania

Abstract. Forecasts of precipitation and air temperature are necessary in order to increase the lead-time of flood warnings and forecasts in a given river basin. The paper presents a coupling procedure for transferring data between the numerical forecast meso-scale meteorological model HRM and the forecast hydrological model CONSUL, which is a rainfall-runoff model. Checking of the results of the coupling procedure between the weather-flood forecast models was performed using the forecasted and recorded hydrometeorological data for some past events in the Crisul Alb and Crisul Negru river basins. This analysis served to assess the use of the coupling procedure in operational flood forecasting and warning.

Keywords: meteorological forecasting model, hydrological forecasting model, coupling procedure.

---

[1] To whom correspondence should be addressed: Rodica Mic, National Institute of Hydrology and Water Management, 97, Soseaua Bucuresti-Ploiesti, 013686 Bucharest, Romania, E-mail: rodica.mic@hidro.ro

1. Introduction

In order to increase the lead time of flood warning, two models with 3-hour time step, such as the ALADIN model (48-hour lead time) and the HRM model (78-hour lead time), were developed in the program of the Romanian National Meteorological Administration.

Simultaneously with the development of numerical models for weather forecasting, high performance flood simulation models capable of dealing with the lead time increase and producing the desired hydrological forecast precision were created. Thus, at present, two such models, the VIDRA model for flood forecasting and the CONSUL model for flow forecasting, are available at the National Institute of Hydrology and Water Management (NIHWM). Both models can simulate a whole range of flows in different river sections, which may be important in flood defence or other aspects of water management, including operation of systems of reservoirs.

By coupling the CONSUL continuous simulation model serving for hydrological forecasts with the HRM model for meteorological forecasts, a flood forecasting system was created and it should be useful for warning the national water management authorities, both those having a forecasting system but not using the present facilities employing automatic consideration of the meteorological forecasts as well as those lacking any flood forecast system at present.

2. Short description of the HRM weather forecast model

The High Resolution Regional Model (HRM) is a flexible tool for numerical weather forecasting, which was developed by the German Meteorological Service (DWD). Starting on Dec. 1, 1999, the NIHWM has been using the HRM mezzo-scale model, which is based on the new operational chain of DWD numerical forecasts. Figure 1 shows the flowchart of hydrological and meteorological data processing.

The HRM model has an Arakawa C latitude/longitude rotated grid, with 41 x 37 points and a network size of 0.25° (~ 20 km) and 20 layers in hybrid vertical coordinates. The formulation of the lateral limits is due to Davies (1976). The initial and boundary condition data are provided by the GME global model, which runs twice per day at DWD.

The HRM model is integrated in Bucharest twice per day for a forecasting interval up to 78 hours. The integration results contain meteorological fields of the forecasting and diagnostic variables for various domains.

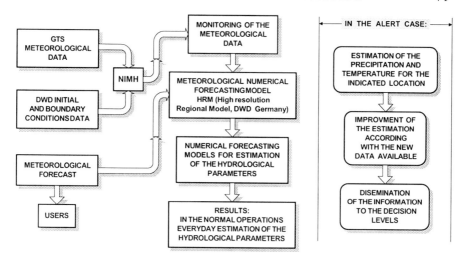

Fig. 1. Flowchart of the HRM model

For hydrological purposes, maps with precipitation accumulated at various time intervals and maps of air temperatures are useful. As an example, maps of 24-hour cumulative precipitation and temperatures at 2-m above the ground, for 12-hour forecasts, are presented in Figures 2 and 3, respectively.

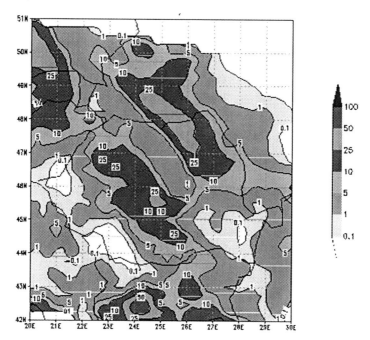

Fig. 2. Map of 24-h cumulative precipitation

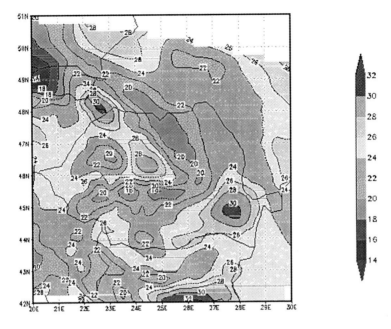

Fig. 3. Map of temperatures at 2 m above the ground

3. Short description of the CONSUL hydrological forecasting model

The CONSUL determinist mathematical model allows flow simulations in flood situation in both small basins and large, complex basins with developed infrastructure. For applying this model to large basins, it is necessary to divide them into homogenous sub-basins. The model computes the discharge hydrographs on sub-basins, and their propagation and superimposition on the main river and on the tributaries.

The modelling of the rainfall-runoff process, which takes place in a hydrographical basin, is accomplished by undertaking the following steps (Serban 1984a, b; Serban et al. 1994; Serban and Corbus 1995):

- Dividing the hydrographical basins in sub-basins, depending on the schematics of flow integration;
- Determining, in each sub-basin, the water supplied by snowmelt;
- Computing, in each sub-basin, the medium influx of water provided by rainfall and snowmelt;
- Determining, in each sub-basin, the net rainfall by subtracting, from the medium input of water, the losses through infiltration and evapotranspiration;
- The net rainfall integration on the slopes and in the primary

hydrographical network, producing as a final result the discharge hydrographs occurring in each sub-basin;
- Superimposing flood waves formed in each sub-basin and computing their propagation through the river channel; and,
- Attenuating flood waves by reservoirs operated in a coordinated way.

The deterministic mathematical rainfall-runoff model simulates the majority of the important hydrological processes in the hydrological basin, namely: snowmelt, interception, retention in depressions, evapotranspiration, infiltration, surface flow, interflow, percolation, and baseflow (Figure 4).

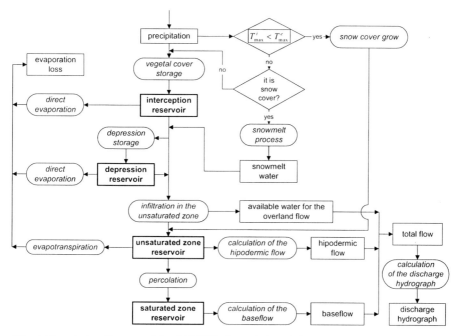

Fig. 4. Descriptive diagram of the CONSUL determinist rainfall-streamflow model

The CONSUL model could be used in flood forecasting, and also for elaborating scenarios in order to determine the response of the hydrological system to severe rainfall impulses, which are rare and did not occur in the past, but could occur in the future. Such scenarios are useful for developing the defence plans for floods.

Depending on the size and stage of development of the infrastructure in the river basin, the following data types are necessary for using the CONSUL model: meteorological and hydrological variables, characteristics of the river basin and also of the hydrographical network.

The meteorological data (precipitation and air temperature averages in the river basins) are provided by the HRM model on a sub-domain of the HTM grid, in every river basin, including the Crisul Alb and Crisul Negru basins.

4. Coupling procedure of the HRM and CONSUL models

The coupling procedure of the HRM and CONSUL models is presented in Figure 5.

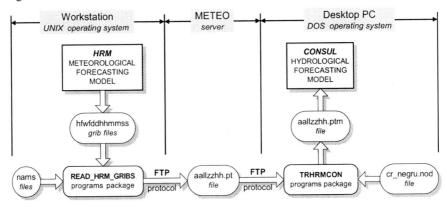

Fig. 5. Coupling procedure of HRM and CONSUL models

For using this procedure it had to be recognised that the HRM model runs on a work station, under the UNIX operation system, while the CONSUL hydrological model runs on a PC under the DOS operation system. Thus, two computer programs were created: READ_HRM_GRIBS that runs on the work station and TRHRMCON that runs on PC.

READ_HRM_GRIBS allows the extraction of the data fields (precipitation and air temperature) forecasted by the HRM in the selected sub-domain of the HRM grid and their writing in 26 ASCII specific files. The names of these files take into account the fact that the data forecasting is made for a period of 78 hours, with a time step of 3 hours, namely *yymmddhh.pt* where *yy* represents the year, *mm* – month, *dd* – the day when the forecasting is elaborated, *hh* - hour, and *pt* – files extensions that are named taking into consideration that files contain precipitation and temperature values. On each line of the files, the precipitation and temperature forecasted values, at that point, are written.

For acquiring the 26 files with forecasted data and the file with geographical coordinates of the HRM grid, the procedures of automatic acquiring by the FTP protocol service were created.

TRHRMCON calculates the precipitation and temperature averages in the sub-basins and accomplishes their transfer into input files specific to the hydrological forecasting model.

The TRHRMCON program is running with the following command line:

>*trhrmcon.exe "nume.cfg"*

where: "*nume.cfg*" represents the link and the name of the configuration file of the application. For the Crisul Alb and Crisul Negru rivers, separate configuration files were created, namely "*cr_alb.cfg*" and "*cr_negru.cfg*".

In the configuration file, the folders should be written with the "*.pt*" type files provided by the HRM model and the input files of the "*.pm*" and "*.tm*" types, which are specific to the discharges in the forecasting model.

TRHRMCON program uses the "*yymmddhh.pt*" files (described above) and "*cr_negru.nod*" files as input data, and calculates the precipitation and temperature averages in sub-basins and writes them in the "*yymmddhh.ptm*" file. Then, from these files the values are inserted into input files specific to the discharge forecasting model, namely "*yy_mm_pm*" and "*yy_mm_tm*", where *yy* and *mm* have the same meaning as above.

In the configuration file "*cr_negru.nod*", the nodes of the HRM grid network corresponding to the Crisul Negru river basin are written. These nodes can be determined using a methodology based on the development of the digital maps of the Crisul Alb and Crisul Negru river basins, together with their sub-basins. The superposition of these digital maps on the meteorological data obtained with the HRM model is made by georeferencing.

Finally, TRHRMCON writes average precipitation and temperature values in each sub-basin considered in the flood forecasting model for the Crisul Negru river basin.

5. Testing of the coupling procedure

The coupling of the CONSUL and HRM models was tested for producing a continuous forecast of the flows in the Crisul Alb and Crisul Negru river basins. To illustrating this process in this paper, the test results for the Crisul Negru river basin are presented. The computation scheme for the flow and its routing is presented in Figure 6.

In Figure 7 the discharges forecasted at four times of the forecast are presented, together with the flood recorded at the Zerind gauging station in March 2000.

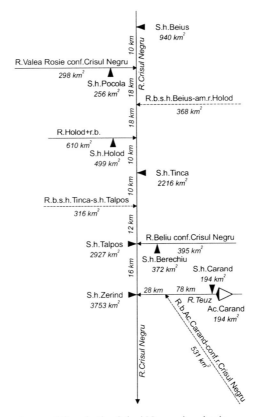

Fig. 6. Computation scheme of flow in the Crisul Negru river basin

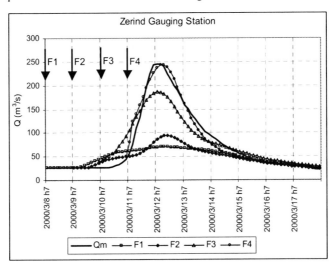

Fig. 7. Comparison between the measured and forecasted discharges at the Zerind gauging station on the Crisul Negru, for four forecast times (F1, ..., F4)

The analysis of the results obtained by coupling the CONSUL and HRM models in the Crisul Alb and Crisul Negru river basins indicates that the procedure is well applicable, especially for the floods generated by spatially uniformly distributed precipitation, as well as in the situations when the physical and geographical conditions of the hydrographical basins will not be modified by natural or anthropogenic causes.

6. Conclusions

The tests demonstrated the reliability of the flood forecasting system in the Crisul Alb and Crisul Negru river basins obtained by coupling the CONSUL hydrological forecasting model with the HRM meteorological forecasting model. Such a development was necessary for modernising the operative forecasting activity and improving the quality of the forecasts (the enhancement of the precision and lead times).

The quantitative forecasting of the hydrological regime will allow a rational use of the water resources, in order to satisfy the users' water demands, and to mitigate flood damages.

In the future, the produced forecasting system could be coupled with the numerical hydraulic model UNDA, which simulates flood routing through river channels, in a natural state or with a developed infrastructure, on the basis of numerical integration of the momentum and continuity equations governing such a flow. For this purpose, one needs to know the geometry of the flow cross-sections and the initial and boundary conditions.

Thus, the development of the HRM-CONSUL-UNDA software package will allow producing flood maps, showing the areas vulnerable to flooding, by using GIS techniques.

References

Davies HC (1976) A lateral boundary formulation for multi-level prediction models. Quart. J. R. Meteor. Soc. 102:405-418

Serban P (1984a) Conceptual model for the determination of the unit instantaneous hydrograph. Hidrotehnica, No. 2, Bucharest

Serban P (1984b) Mathematical models for the flood waves forecast in hydraulic structured basins. Hydrological Studies, Vol. 51, IMH (ed), Bucharest

Serban P, Simota M, Corbus C (1994) Runoff simulation model in the mountain hydrographic basins. International Conference Developments in Hydrology of Mountainous Areas, 12-16 September, Stará Lesná, Slovakia, pp 235-239

Serban P, Corbus C (1995) Le modèle CONSUL de prévision continue des débits. Troisième Rencontres Hydrologiques Franco-Roumaine, 6-8 septembre, Montpellier, pp 155-161

ROUTING OF NUMERICAL WEATHER PREDICTIONS THROUGH A RAINFALL-RUNOFF MODEL

KAMILA HLAVCOVA[1], JÁN SZOLGAY, RICHARD KUBES, SILVIA KOHNOVA, MARCEL ZVOLENSKÝ
Department of Land and Water Resources Management, Slovak University of Technology, Radlinskeho 11, 813 68 Bratislava, Slovakia

Abstract. The applicability of medium range quantitative precipitation forecasts is explored in a flood forecasting system for a medium-size mountainous basin. The results were obtained within the project of the 5th Framework Programme of the European Commission called "European Flood Forecasting System" (EFFS). As a pilot region for the Slovak part of the project, the upper Hron River basin with a drainage area of 1,766 km² was chosen. The basin is located in Central Slovakia and was considered to be representative for mountainous regions where flood generation from cyclonic rainfall and snowmelt processes plays an important role. Meteorological forecasts provided by the European Centre for Medium Range Weather Forecast (ECMWF deterministic model and ensemble forecasts), the Danish Meteorological Institute (DMI – HIRLAM model), the German Weather Service (DWD LM and GME models), and the ALADIN model were used to drive a hydrological model. A conceptual semi-distributed rainfall-runoff model developed at the Slovak University of Technology in Bratislava was used for modelling runoff. The model was calibrated and verified using data from the period of 1991-2000. Hindcasted flows for the floods, which occurred in the upper Hron river basin in July

---

[1] To whom correspondence should be addressed. Kamila Hlavcova, Dept. of Land and Water Resources Management, Slovak University of Technology, Radlinskeho 11, 813 68 Bratislava, Slovakia; e-mail: hlavcova@svf.stuba.sk

1997 and August 2002, were compared with measured flows and further discussed.

Keywords: flood forecasting system, numerical weather forecasts, rainfall-runoff model, flood warning process

1. Introduction

Extreme flood events can have severe consequences for human society. The occurrence of several major flood events in Europe and worldwide over the past decade has also led to concerns about the levels of reliability provided by existing flood forecasting and warning systems. There is a pressing need to re-evaluate existing practices and increase the lead times of forecasts and the reliability of flood warnings. To extend the lead time between the warning and the occurrence of a flood event, advanced methods for runoff and flood prediction based on coupling atmospheric and hydrological models are being developed.

Standardised procedures do not exist in this field, but the application of hydrological modelling in flood forecasting and warning systems is becoming more frequent. Technological advances have led to the development of more complex flood forecasting systems, utilising a variety of statistical models, lumped conceptual hydrological models and hydrodynamic models. Coupled meteo-hydrological model experiments were focused on the Alpine Lago Maggiore basin (Jasper et al. 2002) using the distributed catchment model WaSiM-ETH. 1D and 2D models of flood hydraulics (HEC-RAS, LISFLOOD and TELEMAC-2D) were tested on a 60 km reach of the Severn river in UK, where synoptic images of flooding extent from radar remote sensing satellites were acquired for flood events in 1998 and 2000 (Horritt and Bates 2002). The use of hydrological models for flood simulation is also increasing rather quickly in the pilot region of this study, especially in the retrospective analysis of the generation of extreme events (e.g., Miklánek et al. 2003; Hlavčová et al. 2005).

Within the 5th Framework Programme of the European Commission, a shared cost activity called European Flood Forecasting System (EFFS) has been funded. The goal of the project was to setup and test, in a semi-operational mode, a continental-scale flood forecasting system for major river basins in Europe. The principal research aim was to explore the possibility to extend the lead time of the flood warning process up to 10 days into the future. This was to be achieved by using numerical weather forecasts. Various deterministic and ensemble forecasts delivered by national and international meteorological services have been used within the project to drive a sequence of hydrological rainfall-runoff models and hydraulic models for principal river systems or selected pilot basins. The weather forecasts were downscaled from global circulation models to high resolution local models (Gouweleeuw et al. 2004). The raster based hydrological model LISFLOOD (De Roo et al. 2000) and other models, such as HBV (Bergström 1976), TOPKAPI (Ciarapica and Todini 2002), the FloRIJN model system consisting of the HBV model and the hydrodynamic SOBEK model (Sprokkereef et al. 2001), the Hron model (Szolgay 2003), the NAM model, the VIDRA model and the GAPI-TAPI system (Balint and Szlávik 2000), have been implemented and tested for specific basins.

2. Description of the Upper Hron pilot basin

The upper Hron River basin with a drainage area of 1,766 km^2 was selected as a pilot basin for the Slovak contribution to the EFFS project. The basin is located in Central Slovakia and is representative of high mountainous regions in which runoff generation from rainfall of the cyclonic origin and snowmelt processes plays an important role.

The location of the upper Hron basin in Slovakia with an outlet in Banská Bystrica is presented in Fig. 1. The minimum elevation of the basin is 340 m a.s.l.; the maximum elevation is 2,008 m a.s.l., and the mean elevation is 850 m a.s.l. Seventy percent of the basin area is covered by forests, 10% by grasslands, 17% by agricultural land and 3% by urban areas. A digital elevation model of the basin is shown in Fig. 2.

Flood forecasting in the Hron basin is complex and characteristic for the mountainous regions of the country. In the high Alpine mountain regions, flash floods represent a threat to local villages in narrow valleys. The concentrated character of floods from rainfalls of a cyclonic origin represents the chief danger to major cities and industrial areas with heavy and chemical industries, and electric and nuclear power plants in the central and lowest parts of the catchment. The towns of Brezno, Podbrezová and Banská Bystrica in the catchment are particularly affected. In the lower part of the catchment, floods generated in the upper parts threaten arable land and industrial facilities built along the river. The occurrence of floods seems to be more frequent recently and flood damages are increasing. Flood forecasting is therefore undergoing a major revision. A thorough analysis of the climatic driving factors and the development of new modelling approaches are needed in order to cope with floods.

The present forecasting practice is based on the information from the ALADIN limited area model, near real-time meteorological radar reflectometry data, rainfall depth measurements from 6 telemetric raingauges and a simple forecasting scheme based on travel times inferred from discharge data. Forecasts with lead times up to 24 hours are issued for the main gauge stations. Until now no mathematical rainfall-runoff models and quantitative precipitation forecasting procedures have been used in operational practise. Major stakeholders, including major municipalities, operators of power plants and industry, could greatly benefit from extending the forecast lead time to 3-10 days. For this reason the potential of using middle range meteorological forecasts in hydrological forecasting was tested in the framework of the EFFS project.

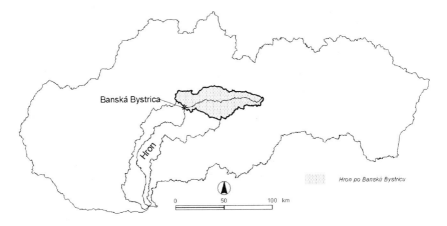

Fig. 1. Location of the Upper Hron River basin (with the Banská Bystrica outlet) in Slovakia

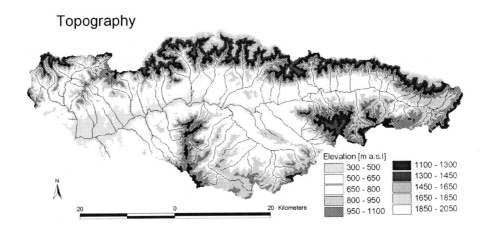

Fig. 2. Digital elevation model of the Hron River basin

3. Description of the Hron model

The Hron rainfall-runoff model developed at the Slovak University of Technology in Bratislava was used for simulating runoff (Szolgay 2003). This conceptual semi-distributed model is based on the HBV model (Bergström 1976), uses daily time steps, and consists of three basic components with 15 calibrated parameters. A scheme of the model is shown in Fig. 3.

The submodel for simulating soil moisture contains 4 parameters and calculates soil water storage, groundwater storage and actual evapotranspiration from the soil profile depending on the relation between the water content in the soil profile *VP* (mm) and the field capacity value *PK* (mm). The actual evaporation from the soil is calculated using linear equation (1) until the ratio *VP/PK* equals or is smaller than the limit of the potential evapotranspiration *LPE*. If *VP/PK* is larger than *LPE*, the actual evapotranspiration *AEv* equals the potential evapotranspiration *PEv*.

$$AEv = PEv\left(\frac{VP}{PK.LPE}\right) \qquad (1)$$

The groundwater recharge is calculated as

$$rech = \left(\frac{VP}{PK}\right)^{RK}.Z(t) \qquad (2)$$

where *RK* is the recharge coefficient, and *Z(t)* is the precipitation and snowmelt [mm].

A runoff submodel with six parameters consists of one nonlinear and one linear reservoir and simulates both quick and slow runoff components (surface and subsurface runoff and baseflow). The runoff is calculated as the sum of all partial runoff components and is routed by a discrete cascade of linear reservoirs. Input data needed for simulating runoff are the catchment average daily precipitation, average mean daily air temperatures, the long-term mean monthly potential evapotranspiration and the long-term mean monthly air temperatures. It is also possible to use daily potential evaporation values if they are available. For the calibration of the model parameters, the observed mean daily discharge values at the outlet cross-section of the selected basin are needed.

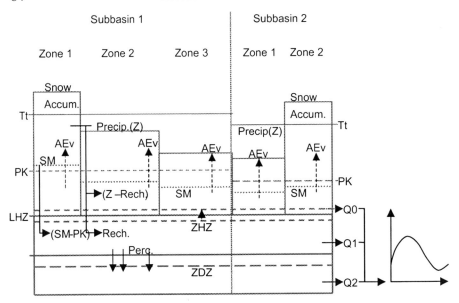

Fig. 3. Schematic representation of the Hron rainfall-runoff model

Legend:
- Tt – threshold temperature [°C]
- Z – precipitation [mm]
- AEv – actual evapotranspiration [mm]
- $Rech$ – recharge [mm]
- SM – soil moisture [mm]
- PK – field capacity [mm]
- LHZ – upper zone limit [mm]
- ZHZ – water storage in upper zone [mm]
- ZDZ – water storage in lower zone [mm]
- $Perc.$ – percolation [mm/day]
- Q_0, Q_1, Q_2 – runoff components [mm]

The Hron model has a built-in calibration procedure using a genetic evolution algorithm. The following four model performance criteria can be alternatively used: the Nash-Sutcliffe criterion, the mean daily error, the mean absolute error and the first lag autocorrelation criterion of the simulation error.

The following input data from the period of 1991-2000 were used for the rainfall-runoff model calibration and validation: basin average daily precipitation from 23 raingauge stations, basin average mean daily air temperature, basin average long-term mean potential evapotranspiration from 6 meteorological stations, and mean daily discharges in the Banská Bystrica outlet. The model was calibrated on the data from the period of 1991-1994; good agreement between the simulated and observed mean daily discharges, and especially peak discharges, was achieved (the

Nash-Sutcliffe criterion had a value of 0.87 and a cumulative difference between the simulated and measured runoff of −4.8 mm/year). Comparison of simulated and measured mean daily discharges for the calibration period is shown in Fig. 4.

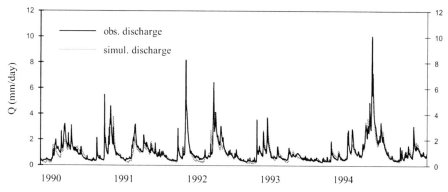

Fig. 4. Comparison of observed and simulated mean daily discharges for the calibration period of 1991-1994

The model was validated for the period of 1995-2000, with an acceptable agreement between the simulated and observed mean daily discharges and peak discharges. The fitness of the model for runoff simulation in the pilot basin was confirmed (Nash-Sutcliffe criterion of 0.76 and a cumulative difference between the simulated and measured runoff of −35.8 mm/year can be considered as adequate). The decrease in the values of the model performance criteria is attributed to the spatial allocation of rainfall stations in the high mountainous part of the catchment. These should be substantially upgraded in the near future, so the model performance can be expected to improve also.

4. Numerical weather prediction models

Large scale weather forecasts for the EFFS were derived using the prediction system developed by the European Centre for Medium-Range Weather Forecasts (ECMWF) and the global scale hydrostatic grid-point model (DWD-GME) developed by the German Weather Service (DWD). The vertical resolution of the ECMWF system is represented by 60 layers up to 0.1 hPa, and the horizontal resolution is approximately 40 km. In the ensemble mode the vertical resolution is reduced to 40 vertical levels and the horizontal resolution to 80 km. In both modes, the system produces a forecast of weather variables for each grid cell every 6 hours for up to 10 days ahead. The DWD-GME model works with a horizontal resolution of about 60 km and produces forecasts every 6 hours for up to 156 hours ahead.

To increase the forecast spatial and temporal resolutions, regional Numerical Weather Prediction models with lead times of up 3 days were used to downscale the output from global NWP models: the DMI version of the High Resolution Local Area Model (HIRLAM) developed by the Danish Meteorological Institute, the DWD Local Model (DWD-LM) and the ALADIN Lace model.

For the Hron River basin, quantitative precipitation forecasts were provided by all the above listed NWP systems in relevant grid points of the basin for two extreme meteorological events, which occurred in July 1997 and in August 2002. Forecasted and observed basin average daily precipitations were compared and analysed. As an example, comparison of observed basin average daily and accumulated precipitation with 18-114 hour forecasts provided by the ECMWF deterministic model for August 2002 is shown in Fig. 5 and comparison with 18-66 hour forecasts provided by the DMI HIRLAM model is shown in Fig. 6.

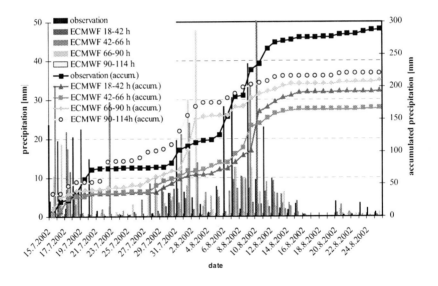

Fig. 5. Observed and forecasted precipitations for the period from July 15 to Aug. 26, 2002 (forecasts were produced 18-114 hours ahead by the deterministic ECMWF model)

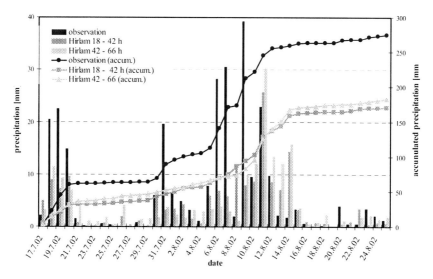

Fig. 6. Observed and forecasted precipitations for the period from July 15 to Aug. 26, 2002 (forecasts were produced 18-66 hours ahead by the DMI HIRLAM model)

It can be concluded from these comparisons, that all precipitation forecasts gave comparable results and, a relatively good timing was achieved, especially for 1-2 days forecasts. Generally, for daily precipitation better results were achieved by the regional NWP systems with shorter lead time than by global models. But observed cumulative precipitation was higher than forecasts in all cases; the best results with respect to cumulative precipitation were achieved with the DWD-GME model and the worst with the ALADINE model.

5. Application to the 1997 and 2002 Hron River floods

Flood hindcasts based on the NWP forecasts were simulated using the Hron hydrological model for two selected historical flood events in the upper Hron basin: ECMWF deterministic forecast for the July 1997 flood, ECMWF deterministic and ensemble forecasts for the August 2002 flood (Fig. 7), and DWD GME, DWD LM, DMI Hirlam, and ALADIN forecasts for the August 2002 flood (Fig. 8).

For the Hron River basin one and two day ahead flood forecasts using the DWD, DMI and ECMWF models produced acceptable results, as did five-day ahead flood forecasts using DWD-GME (Figs. 9 and 10). However, it was not possible to generalise the results on the basis of forecasts for only two flood events.

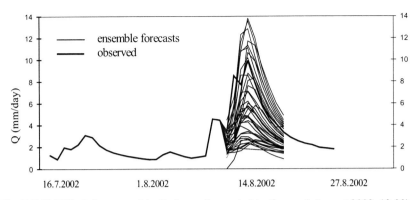

Fig. 7. ECMWF: 9-day ensemble discharge forecasts (starting on 9 August 2002, 12:00)

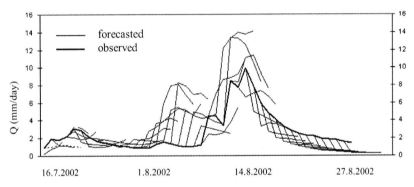

Fig. 8. DWD GME: 5-day discharge forecasts (August 2002)

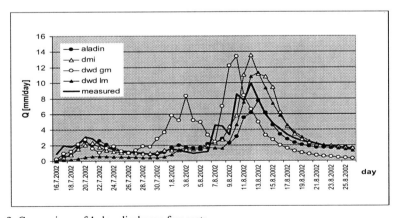

Fig. 9. Comparison of 1-day discharge forecasts

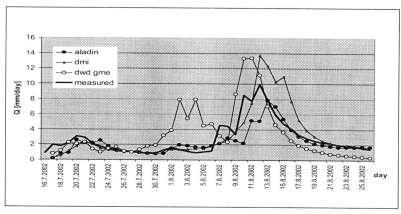

Fig. 10. Comparison of 2-day discharge forecasts

The comparison of forecasted and observed floods indicates the following: a relatively good timing of forecasted and observed flood peaks was achieved, flood peaks forecasted by the DWD, ECMWF and DMI models were overestimated, and the flood peaks forecasted by using the ALADIN meteorological forecasts were underestimated compared to the observed floods.

6. Conclusions

A good cooperation between hydrologists and meteorologists was achieved during the EFFS project in Slovakia. The information gained by using one to two day quantitative precipitation forecasts from different LAMs was usefully supplemented by the ensemble forecasts. In general it can be stated that the benefit of using ensemble forecasts has been proven also for mid-size catchments with a rather sparse coverage of meteorological model grid points. It was demonstrated that in some cases both deterministic and ensemble forecasts gave a clear flood signal up to 4 days in advance, but there was a considerable variability in the forecasts, which would have to be reduced in the future.

The analysis of longer time series would have been needed in order to adequately address uncertainty and usefulness of the ensembles. Also ways how to meaningfully interpret the ensembles and communicate the information to the users are not yet fully established. It is felt that end users should be trained how to interpret ensemble and probabilistic forecasts. The response of water authorities, civil protection authorities and related services was also not tested or practiced in the project. Discussions with end users have shown that the fact that drought situations were not explicitly handled in the model tests, can be also considered as a shortcoming.

Acknowledgement

The authors gratefully acknowledge the support from the VEGA grant agency in the form of grants 1/1145/04, 1/2032/05 and 2/5056/25, and the European Commission for the support of the EFFS.

References

Bálint G, Szlávik L (2000) Flash floods, their analysis and frequency (Learning of case studies in Nothern Hungary). In: Proc. 20th Conference of the Danubian Countries. SVH, SHMI, Bratislava

Bergström S (1976) Development and application of a conceptual runoff model for Scandinavian catchments. Swedish Meteorological and Hydrological Institute, Report RHO, No. 7, Norrköping, Sweden

De Roo A, Wesseling CG, Van Deursen WPA (2000) Physically based river basin modeling within a GIS: The LISFLOOD model. Hydrological Processes 14:1981-1992

Ciarapica L, Todini E (2002) TOPKAPI: A model for the representation of the rainfall-runoff process at different scales. Hydrological Processes 16:27-229

Gouweleeuw B, Reggiani P, De Roo A (2004) A European flood forecasting system EFFS. Final Report, JRC, WL Delft Hydraulics, Italy

Hlavčová K, Kohnová S, Kubeš R, Szolgay J, Zvolenský M (2005) Estimation of future flood risk for flood warning systems. Hydrology and Earth System Sciences (Special Issue), 9:431-448

Horritt MS, Bates PD (2002) Evaluation of 1D and 2D numerical models for predicting river flood inundation. Journal of Hydrology 268(1-4):87-99

Jasper K, Gurtz J, Lang H (2002) Advanced flood forecasting in Alpine watersheds by coupling meteorological observation with distributed hydrological model. Journal of Hydrology 267: 40-52

Miklánek P, Halmová D, Pekárová P (2000) Extreme runoff simulation in Mala Svinka Basin. Conference on Monitoring and Modelling Catchment Water Quality and Quantity, Ghent University, Belgium, pp 49-52

Sprokkereef E, Buiteveld H, Eberle M, Kwadijk J (2001) Extension of the flood forecasting model FloRIJN. NCR Publication 12-2001, ISSN No. 1568234X

Szolgay J, Hlavčová K, Kubeš R, Kohnová S (2003) Medium range forecasts in the Hron River Basin. In: Proc. 3rd Water Management Conference. ÚVS FAST VUT, Brno 2003, pp 226-235

THEORETICAL GROUND OF NORMATIVE BASE FOR CALCULATION OF THE CHARACTERISTICS OF THE MAXIMUM RUNOFF AND ITS PRACTICAL REALISATION

EUGENE GOPCHENKO[1], VALERIYA OVCHARUK
Odessa State Environmental University, Lvovskaya15, 65016 Odessa, Ukraine

Abstract. A new mathematical model for calculation of the maximum runoff is proposed, considering precipitation input, overland flow, storage in lakes and reservoirs, and streamflow. Examples of the model application to Ukrainian rivers are given.

Keywords: Maximum runoff, floods, standardisation of flow characteristics

1. Introduction

The methods used to calculate maximum runoff characteristics can be classified into the following groups: (a) Limiting intensity formulas (like the Rational Method), (b) volumetric formulas, (c) empirical reduction formulas, (d) the estimated limiting storm methods (e.g., the probable maximum storm), and (e) formulas based on the theories of bed isochrones. In the first case, an empirical dependence between the maximum intensity of rainfall for a certain time interval is established; however, without a precise physical substantiation. Volumetric and reduction formulas assume uni-modal floods, with the volumetric structure providing the only general solution. The reduction formula represents only a special case of such a

---
[1] To whom correspondence should be addressed. Eugene Gopchenko, Odessa State Environmental University, Hydrology Department, Lvovskaya15, 65016 Odessa, Ukraine; e-mail: gidro@ogmi.farlep.odessa.ua

solution, corresponding to the case, when the duration of overland flow into the channel network changes insignificantly throughout an area. To establish a volumetric structure for setting of the characteristics of the maximum runoff is rather difficult, since the flood duration depends not only on the size of the catchment area, but also on factors controlling overland flow (under the influence of forests and swamps) and the effects of channel and flood plain storage. For this reason, the volumetric-type formula did not gain much ground on a global scale. On the contrary, simple techniques of data generalisation according to the reduction formulas and the methodology of 'the upper limit' have dominated the modern developments in theoretical evaluation of characteristics of maximum runoff and brought about stagnation in this field for many years.

2. A proposed model of maximum flow

The authors proposed a variant of an analytical scheme describing the basic model of runoff generation as shown in Fig. 1. By considering Fig. 1, it is easy to conclude that the idea of developing limiting intensity formulae (the rational method) is theoretically inconsistent, because in the rainfall-streamflow model an operator for transformation of precipitation into overland runoff is missing. It is obvious that the process of streamflow formation should be described by two operators referring to: (a) the rainfall-overland flow, and (b) overland flow-streamflow. This reasoning should form the theoretical and methodical bases for solutions of problems related to, for example, forecasting of stream hydrographs on the basis of atmospheric precipitation data. However, in hydrological practice, it is not the whole channel hydrograph that is of profound interest when looking for solutions to diverse tasks, but only its maximum ordinate, corresponding to an infrequent exceedance probability.

In this case, the theoretical basis for design can be simplified significantly, by restricting consideration to the overland flow-streamflow operator, which can be described by the balance equation (Bephani et al. 1981):

$$V \frac{\partial Q}{\partial x} + \frac{1}{\varepsilon_F} \frac{\partial \omega}{\partial t} = q'_t B_t, \qquad (1)$$

where Q = water discharge; ε_F = coefficient of channel and flood plain storage; w = total area of the flow cross-section in the plane of channel time lag isochrones; q = arbitrary function of overland inflow into the channel network; and, Bt = width of a watershed corresponding to the channel lag isochrones at time t.

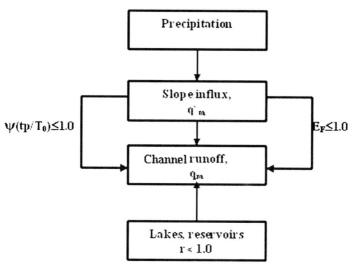

Fig. 1. Scheme of streamflow generation

If in equation (1) one is interested only in the maximum discharge Q_m, then according to Gopchenko and Ovcharuk (2002), functions q'_t and B_t could be expressed from the maximum value as

$$q'_t = q'_m \left[1 - \left(\frac{t}{T_0}\right)^n\right], \qquad (2)$$

and

$$B_t = B_m \left[1 - \left(\frac{t}{t_p}\right)^m\right]. \qquad (3)$$

Where q'_m = maximum specific discharge of overland flow; B_m - maximum width by the channel lag isochrones; T_0 = duration of the overland flow; and t_p = time lag of the flood wave.

By expressing the water discharge Q_m as a product of the flood wave celerity V and the flow cross-sectional area w, and substituting eqs.(2) and (3) into eq.(1), we shall derive:

$$V\frac{\partial \omega}{\partial x} + \frac{1}{\varepsilon_F}\frac{\partial \omega}{\partial t} = q'_m B_m \left[1 - \left(\frac{t}{t_p}\right)^m - \left(\frac{t}{T_0}\right)^n + \frac{t^{n+m}}{T_0^n t_p^m}\right]. \qquad (4)$$

The integration (4) provided that $t_p < T_0$, results in expression

$$q_m = q'_m \varepsilon_F \left[1 - \frac{m+1}{(n+1)(m+n+1)} \left(\frac{t_p}{T_0} \right)^n \right]. \tag{5}$$

When $t_p \geq T_0$,

$$q_m = q'_m \varepsilon_F \frac{T_0}{t_p} \frac{n}{n+1} \left[\frac{m+1}{m} - \frac{n+1}{m(m+n+1)} \left(\frac{T_0}{t_p} \right)^m \right]. \tag{6}$$

From eqs. (5) and (6) it is apparent, that the specific discharge of overland flow and coefficients of channel and flood plain storage are included in the right-hand side of equations, as well as the transformation functions of $\psi(t_p/T_0)$, which can be expressed as:

a) for $t_p < T_0$

$$\psi\left(\frac{t_p}{T_0}\right) = 1 - \frac{m+1}{(n+1)(m+n+1)} \left(\frac{t_p}{T_0} \right)^n ; \tag{7}$$

b) for $t_p \geq T_0$

$$\psi\left(\frac{t_p}{T_0}\right) = \frac{n}{n+1} \frac{T_0}{t_p} \left[\frac{m+1}{m} - \frac{n+1}{m(m+n+1)} \left(\frac{T_0}{t_p} \right)^m \right]. \tag{8}$$

By combining eqs. (5) and (6), and considering eqs. (7) and (8), it is possible to produce a generalised equation for q_m:

$$q_m = q'_m \psi(t_p/T_0) \varepsilon_F . \tag{9}$$

If there are lakes or reservoirs through which the river is passing, lake (or reservoir) control coefficient must be introduced on the right-hand side of eq. (9), i.e.

$$q_m = q'_m \psi(t_p/T_0) \varepsilon_F \cdot r \tag{10}$$

In eq. (10) the maximum specific discharge of overland flow proved to be a rather informative parameter which on the basis of integration (2) is equal to

$$q'_m = \frac{n+1}{n} \frac{1}{T_0} Y_m , \tag{11}$$

where $n+1/n=$ irregularity coefficient for overland flow, and $Y_m =$ the area of the overland flow.

3. Analysis of the equation parameters

It is obvious, that q_m represents a complex characteristic, which could be related to other parameters of the overland flow hydrograph; its shape - through $n+1/n$, volume - through Y_m, and duration of inflow through - T_0. On the other hand, being the characteristics of overland flow, T_0 and Y_m may account for controlling influence of such local physiographic and hydrological factors, as the forest cover or inundation, i.e. $T_0(f_f, f_s)$ and $Y_m(f_f, f_s)$.

Thus,

$$q'_m = \frac{n+1}{n} \frac{1}{T_0(f_f, f_s)} Y_m(f_f, f_s), \qquad (12)$$

As an example for floods on the Transcarpathian rivers in the cold period the following expressions are derived:

$$T_0(f_f) = 142 + 0.075 f_f \qquad (13)$$
$$Y_m(f_f) = 203 - 0.42 f_f \qquad (14)$$

Using eqs. (13) and (14) in conjunction with eq. (12), it is rather simple to model q_m and thereby determine not only the direction, but the extent of influence of the forest cover on the maximum streamflow of the Transcarpathian rivers.

The dependence $q_m = f(f_f)$ expressed in relative coordinates is presented in Fig. 2. The diagram has rather important practical significance, because it describes the essential regulating influence of the Transcarpathian forests on overland flow of high waters.

The upper physical limit of function $\psi(t_p/T_0)$ corresponds to $t_p/T_0 = 0$, and the lower one is zero (at $t_p \ll T_0$). The visualisation of the role of channel lag in transformation of the overland flow can be best achieved with the help of Fig. 3.

As it can be seen, even at $t_p/T_0 = 1.0$, $\psi(t_p/T_0) = 0.25$ (at n = 0.2) and 0.50 (at n = 1.0). It should be noted that in most cases spring floods (snowmelt driven) are characterised by 'n' values of about 0.1 - 0.2, whereas the floods of storm origin are formed at n = 0.5 - 1.0. The value of the coefficient for channel and flood plain storage ε_F can be determined from eq. (9), as

$$\varepsilon_F = \left(q_m / q'_m\right) / \psi\left(t_p / T_0\right) \tag{15}$$

Where q_m / q'_m = coefficient of general reduction varying from 1.0 (at $F \to 0$) to 0 (at $F \to \infty$).

As the overland flow duration varies slightly in space, the T_0 coefficient of general reduction is

$$q_m / q'_m = \frac{1}{(F+1)^{n_1}} \tag{16}$$

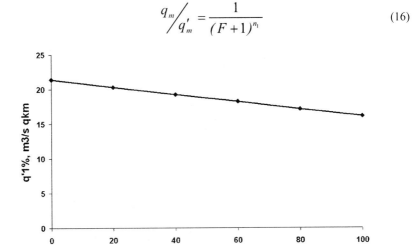

Fig. 2. Dependence of the maximum specific discharge of overland flow on forest cover (%)

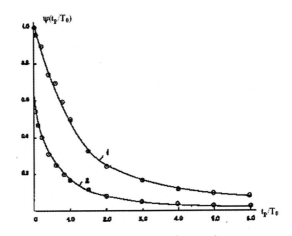

Fig. 3. Dependence of the transformation function $\psi\left(t_p / T_0\right)$ on ratio t_p/T_0: (1) n = 1.0, (2) n = 0.2

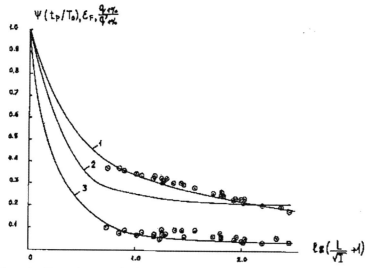

Fig. 4. Dependence of the transformation coefficients on the watershed area: (1) $\psi(t_p/T_0)$, (2) ε_F; (3) $q_{1\%}/q'_{1\%}$

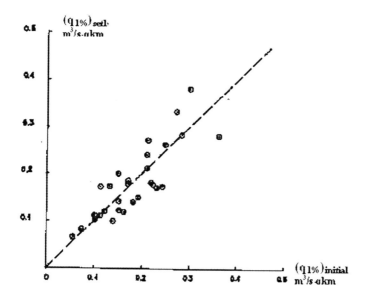

Fig. 5. Comparison of specific discharges $(q1\%)_s$ with the initial data $(q1\%)_i$

Then

$$\varepsilon_F = \frac{1}{(F+1)^{n_1}} \bigg/ \psi(t_p/T_0) \qquad (17)$$

In computations, relation q_m/q'_m can be tabulated or represented as a function of the catchment size. The above discussion offers a conclusion that for a known duration of overland flow T_0 there would be no problems to apply the operator described by eq. (9) or (10). Unfortunately, in stream gauging network measurements, overland flow characteristics are not measured. Therefore, a method for numerical solution for T_0 in eq. (9), as well as the one accounting for eqs. (7) and (8) was developed by the Chair of the Hydrology of the Land Department at the Odessa State Environmental University.

Some general ideas for general reduction value q_m/q'_m for a spring flood with a 0.01 exceedance probability for the rivers of the Sea of Azov basin and its components - $\psi(t_p/T_0)$ and ε_F can be found in Fig. 4. The figure shows that the share of channel attenuation of flood waves and the influence of channel and flood plain storage in transformation of maximum discharges are almost identical. The application of eq. (9) was executed by the authors on an example of rivers flowing into the Sea of Azov. The comparative analysis of the results is given in Fig. 5.

4. Conclusions

The existing methods of standardisation of maximum runoff characteristics used in various countries do not always differ with respect to their theoretical validity and optimality of basic structures. The technique proposed by the authors scientifically and formally transforms overland flow into streamflow. The basic structure differs by universality both in the range of catchment areas (it is applicable practically to the whole range from separate slopes to large river systems) and genetic types of floods. Transformation factors $\psi(t_p/T_0)$ and ε_F, confined by the upper limit, make it possible to carry out reliable extrapolation of dependencies $\psi(t_p/T_0)$ and ε_F for areas of small catchments without observed data.

References

Bephani AN, Bephani NF, Gopchenko ED (1981) Regional models of rain flood formation on the territory of the USSR (in Russian). Obninsk

Gopchenko ED, Ovcharuk VA (2002) Formation of maximum runoff of spring high water under conditions of Southern Ukraine (in Russian). Odessa, TES

SCENARIOS OF FLOOD REGIME CHANGES DUE TO LAND USE CHANGE IN THE HRON RIVER BASIN

ZORA PAPANKOVA[1], OLIVER HORVAT, KAMILA HLAVCOVA, JAN SZOLGAY, SILVIA KOHNOVA
Department of Water and Land Resources Management, Bratislava, Slovak Republic 81368

Abstract. Possible changes in the flood regime of the Hron river basin due to land use change were estimated by using a physically based distributed rainfall- runoff model for modelling runoff from rainfall and snowmelt. Model parameters were estimated from data from the 1981-2000 period and three basic digital map layers: land use map, soil map and digital elevation model. Several scenarios of land use change were prepared and runoff under different land use conditions was simulated. Design maximum mean daily discharges with the return period of 100 years under current and changed land use were estimated and compared.

Keywords: rainfall-runoff modelling, distributed parameters, land use change scenarios, runoff formation, partial runoff components, flood regime

1. Introduction

Land use is defined by Veldkamp and Fresco (1996) as a human activity directly related to land, making use of its resources and interfering with the ecological processes that determine the functioning of land-cover. Modified human or biophysical demands arising from changed natural, economic or

---

[1] To whom correspondence should be addressed. Zora Papankova, Slovak University of Technology, Department of Water and Land Resources Management, SUT, Radlinskeho 11, 813 68 Bratislava, Slovak Republic, e-mail: papankova@svf.stuba.sk.

political conditions have led to extensive land use change (O'Callaghan 1996). The rapidly increasing population pressure in many rural areas has often led to changes in terms of deforestation, reclamation of wetlands, etc. with the aim of increasing agricultural production.

It is generally recognised that in a number of different regions this has resulted in land degradation in terms of soil erosion, reduced productivity and deterioration of fragile natural ecosystems. According to Riebsame et al. (1994) alterations in land-use exert an influence on the whole ecosystem, thus affecting the radiation budgets and water balance.

In the debate among scientists and water resources managers about the possible impacts on the water resources of past and ongoing land use changes, it has become increasingly clear that there is a need for improved knowledge and quantitative documentation of the impact of changes in land use and management practice on land and water resources. The effects of land use change on catchment hydrological responses, especially connected to the forest management have been documented at smaller watershed scales (e.g., see Hibbert 1967; Bosch and Hewlett 1982; Harr 1986; Jones and Grant 1996; Stednick 1996). Removal of forest cover is known to increase streamflow as a result of reduced evapotranspiration and to increase peak flows as a result of higher water tables (Mattheusen et al. 2000).

Physically based rainfall-runoff models are used in hydrology for a wide range of applications such as the extension of streamflow records, estimation of flows for ungauged catchments, prediction of the effects of land use change and examination of the effects of climate change. The impact of land use changes on storm runoff generation is a current topic in hydrologic research and is often assessed by rainfall-runoff model simulations (e.g. Bultot et al. 1990; Parkin et al. 1996). Quite a number of rainfall-runoff models have been developed over the past three decades; typical of these are lumped parameter models and distributed parameter models.

It has been claimed that distributed parameter models are more suited to predict the hydrologic effect of land use change because their parameters have a physical interpretation and the structure allows for an improved representation of spatial variability. However, the current generation of distributed models can be considered as lumped conceptual models because they use equations based on small scale physics, applied at grid scale (Beven 2001). In ideal cases (intensive data collection), model parameters are measured or estimated from catchment characteristics. More commonly, however, distributed models have their parameters determined from calibration, because of the unknown spatial heterogeneity of parameter values and the cost involved in measurements.

In this paper, different scenarios of changed land use conditions for the upper Hron River basin were developed for the purpose of evaluation of the impact of land use changes on the runoff regime. The distributed physically based rainfall runoff model WetSpa was used to simulate total runoff for the prepared scenarios. The results were compared with the runoff under actual land use conditions with emphasis put on the surface runoff partial component responsible largely for the flood regime.

2. Description of the rainfall-runoff model with distributed parameters

The rainfall-runoff model is based on the structure of the physically based model WetSpa (Wang et al. 1996). Its components and model structure were slightly changed in the solution progress so that model was more appropriate for runoff modelling from rainfall and snowmelt in the upper Hron River Basin (Szolgay et al. 2004). This physically based rainfall-runoff model with distributed parameters divides the basin into uniform spatial units on grid scale, in which the hydrological balance and runoff are calculated up to the basin outlet. Individual components of the hydrological balance are liquid and solid precipitation, interception, soil moisture, infiltration, actual evapotranspiration, surface runoff, interflow in the root zone, percolation into the groundwater, groundwater runoff and production of groundwater recharge in the saturated zone. Transformation of the surface runoff in the catchment is simulated by the diffusion wave approximation and using hydraulic characteristics of water flow on hill slopes and in the stream network. The interflow and percolation of each cell is calculated by Darcy's law and the kinematic wave approximation.

The rainfall-runoff model is executed as an ArcView GIS extension and the whole preparation of spatially distributed data is fixed to the GIS interface. Model inputs are prepared as maps in digital form, hydrometeorological data and physiographical properties of the given area are in text format. Hydrological and climatic data for modelling are daily or hourly total precipitation values from raingauge station measurements, mean daily or hourly values of the average temperature from climatological station measurements and mean daily or hourly measured discharges from the river gauging station at the basin outlet. Apart from a large number of physically based parameters, which are derived from physiographical properties of the catchment, the model requires additional input of 11 global parameters, which can be calibrated.

The model was calibrated on the upper Hron River Basin up to the outlet gauging station in Banská Bystrica for selected periods from the whole period from 1981 to 2000. The congruence rate between measured and simulated mean daily discharges was assessed by Nash-Sutcliffe coefficient, the best value obtained for the whole 20-year period was 0.732.

3. Description of the Hron River basin and input data

In this study the Hron River basin, which has an area of 1766 km^2, was selected as a pilot basin. The Hron River basin is located in central Slovakia; the minimum elevation of the basin is 340 m a.s.l.; the maximum elevation is 2,004 m a.s.l.; and the mean elevation is 850 m a.s.l. Seventy percent of the basin area is covered by forest, 10% by grasslands, 17% by agricultural land and 3% by urban areas (the digital elevation model of the basin is shown in Figure 1).

Necessary spatial layers of physiographical characteristics are the digital elevation model, map of land use types and map of soil types. The digital elevation model (DEM) with resolution 20 x 20 m was derived from digitised contour lines of the Basic Map Work of the Slovak Republic (1:10,000) by interpolation. Stream network as a reference grid relief was also generated from BMW SR with special elevation points that were used for improving the quality of valley and alluvial sections topography. Derived layers from DEM represent map of flow accumulation, flow direction, slope map, map of hydraulic radius and delineation into sub-watersheds.

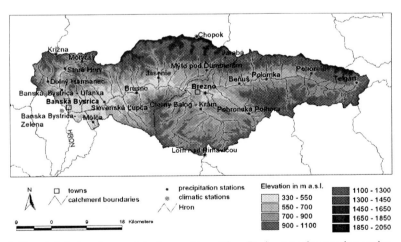

Fig. 1. The digital elevation model of upper Hron River Basin up to the gauging station Banská Bystrica

The land use map originated from thematical mapping of Slovakia by satellite LANDSAT. The images were made in the year 2000 and their resolution is 30 x 30 m. They were corrected by topographical, crop and forest cover maps in the scale 1:10,000. Land use was divided to 16 categories; in the upper Hron River Basin 15 categories can be found that had to transformed to the 14 categories that are defined for WetSpa (Liu and Smedt 2004). Daily total precipitation was measured at 20 raingauge stations, daily average temperature at 6 climatic stations.

4. Land use change scenarios in the Hron River basin

To exploring the changes in land use in the basin, different land use scenarios were created. Designing scenarios is a widespread technique in business, planning and policy consulting, because it offers an opportunity for assessment of the present and possible future situations and the mitigation of mismanagement (Niehoff 2002). The scenarios created for this purpose were selected to account for the impact of the possible changes mainly in forest, farmland and urban area land use types on runoff. Using these scenarios, runoff from rainfall and snowmelt was simulated in daily steps for the period 1981-2000. The resulting changes in runoff were evaluated by comparing the simulated daily average runoff series and their statistical characteristics for the actual land use and the land use scenarios, as well as by plotting the runoff changes spatially in a map. The change in runoff was evaluated also for runoff components – overland flow, subsurface flow, baseflow and total runoff.

The following scenarios were created to account for the land use change:

- Natural land use – scenario representing land use closest to that of a natural, pristine landscape, with almost the whole basin area covered with forest,
- Change in forest composition – a way of changing land use towards the natural state that would be possible while respecting the existing land use i.e. urban land, farm land etc.,
- Grass over forest – scenario suggesting the forest becoming grass land,
- Change in impermeable areas – scenario, where the impermeable areas of urban land make up 100%,
- Grass over farmland – scenario suggesting that arable land be left as grass, and
- Afforestation of critical hydrotopes – scenario, where areas that generate most runoff (according to the combination of various geophysical characteristics defined for the hydrotopes) would be afforested.

4.1. CHANGES IN SURFACE RUNOFF

Simulation and analysis did not address just the balance in the total runoff regime, but also the partial runoff components' contribution to the total change. Graphically, the results of total runoff and runoff components for the different scenarios are plotted in Figure 2. When comparing the partial runoff component results, it is obvious that surface runoff is the component with the most evident changes for the given scenarios. It can be seen that the surface runoff component is predominantly accountable for the total change in runoff, and therefore it is clear that for the purpose of flood regime evaluation for the given scenarios this is the crucial element that must be further explored. The baseflow runoff component is mostly constant for all the scenarios and the only other significant change is in the interflow runoff component for the natural land use scenario, where the almost total extent of forest in the basin results in interflow mean annual runoff depth being down 72 mm/year from the actual state. Summarised in Tables 1 and 2 are the results of the comparison of total runoff changes for different scenarios, where the long-term average annual runoff depths in mm/year are listed for the scenarios and for the actual land use, as well as the difference in runoff in the scenarios vs. actual land use. The surface runoff values for the different scenarios deviating from the value for the reference state of actual land use can be seen in Figure 3.

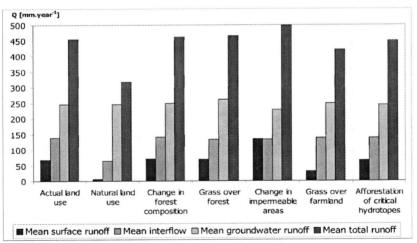

Fig. 2. The comparison of the mean annual depth of the total runoff and its components for the actual land use situation and considered scenarios of land use changes

The comparison between daily average discharge for the scenario "Natural land use" and the existing land use, suggests that the almost complete afforestation of the basin can indeed lead to a very significant decrease in daily runoff. The decrease in runoff is evident also in average monthly values, especially for the months of April, May and June. For the annual values this decrease is also evident, in the long-term daily average discharge as well as in the average annual runoff depth. The average annual runoff depth has decreased by 140 mm/year, representing a difference of -30% with respect to the actual state. The decrease for the surface runoff is 63 mm/year, i.e. -91% compared to the actual state.

Table 1. Mean annual values of the surface runoff depth (mm/year) for the actual land use and land use change scenarios

Land use change scenarios	Mean surface runoff depth (mm.year^{-1})
Actual land use	68.98
Natural land use	5.72
Change in forest composition	71.68
Grass over forest	69.53
Change in impermeable areas	135.27
Grass over farmland	32.35
Afforestation of critical hydrotopes	66.13

Table 2. The change of mean annual depth of the total runoff (mm.year^{-1}) for land use change scenarios vs. the actual land use

Land use change scenarios	Mean surface runoff (mm.year^{-1})
Natural land use	-63.3
Change in forest composition	2.7
Grass over forest	0.6
Change in impermeable areas	66.3
Grass over farmland	-36.6
Afforestation of critical hydrotopes	-2.9

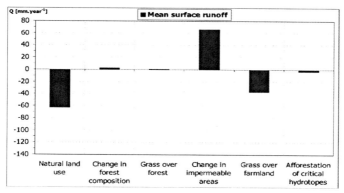

Fig. 3. The comparison of the mean annual depth of the surface runoff (mm.year^{-1}) for land use change scenarios vs. the actual land use

From the results of daily average discharges simulated for the scenario "Change in forest composition" comparison to the actual state it can be seen that the change in forest composition in the Hron basin has little to no effect on the runoff, and that is true for both the monthly and long-term annual average daily discharges. The total change in the average annual runoff depth is 8.3 mm/year, which is insignificant. A similar small change can be observed for the surface runoff component.

The results of daily average discharge for the scenario "Grass over forest" suggest that compared to the actual state, the overall runoff can be slightly higher. This increase can be seen in the daily average discharge as well as in the increase of the average annual runoff depth. From the comparison of the average daily discharge for the individual months it is observed that the increase in daily runoff is visible throughout the whole year; the most evident being from October through December (approx. +10% of the actual state). During the transition between the spring and summer (April - June) a slight decrease (up to -1.5%) in runoff can occur. Nevertheless, on average, an overall increase of runoff can be expected for this scenario; the average annual runoff depth rising by about 12 mm, which represents an increase of +2.5%. Surface runoff changes only little for this scenario. This result can be attributed to the parameterisation of forest and grassland land use types in the model (root depth, interception capacity, roughness) that can affect the process of forming partial runoff components in the model and it can have a different manifestation in various seasons of the year.

The comparison of daily average discharges for the scenario "Change in impermeable areas" to the actual state has shown that changing the entire urban area surface into impermeable areas can cause a daily average discharge increase in the whole area of the basin. The increase in runoff is visible in all months, especially so in the summer period. For the annual values the increase has been observed in the long-term mean of daily average discharge and also in the average annual runoff depth. The average annual runoff depth has increased up to 45 mm, i.e. +10% difference compared to the actual state. There was a substantial increase in the surface runoff – up 66 mm/year, i.e. +96% compared to the actual state.

The difference in daily average discharge for the scenario "Grass over farmland" compared to the actual state means that for this scenario a decrease in runoff is to be expected. The decrease is apparent in monthly values of daily average discharge for every month and the year as whole as well as in the smaller average annual runoff depth. The total average runoff depth has decreased 32 mm, i.e. -7% decrease compared to the actual state. The total decrease has been caused primarily by the decrease in surface runoff, its average runoff depth going down 37 mm (-53% of the actual state).

From the comparison of the daily average discharge for the scenario "Afforestation of critical hydrotopes" and the actual state, it can be observed that almost no change in runoff occurred for this scenario. The total average annual runoff depth was 5 mm down from the actual state, accounting for -1% change compared to the actual state. These results are probably caused by the critical areas in the Hron basin (steep slopes, less-permeable soils) being actually largely forested and therefore making the theoretical scenario for this basin close to the actual state.

4.2. CHANGES IN DESIGN FLOODS

In order to account for the flood regime changes due to the land use changes in the simulated scenarios, a design flood analysis was undertaken. For each land use scenario and for the actual state design floods for 2-, 5-, 10-, 20-, 25-, 50-, 100-year return periods were calculated. In this study, the DVWK/101 (1999) method used in Germany was applied for estimation of design values of maximum discharges. DVWK/101 (1999) describes the methodology for at-site estimation of design values from annual and seasonal maximum data and is supported by the HQ-EX statistical computer program (version 2.02, 1997) developed by Wasy Gmbh.

The plotting positions of the maximum precipitation totals are calculated according to Cunnane, WMO (1989) as:

$$P = \frac{m - 0.4}{n + 0.2} \qquad (1)$$

where n is the sample size and m is the rank of the observations in descending order.

To estimate the parameters of theoretical distribution functions three methods can be used alternatively:

- the method of moments (MOM),
- the maximum likelihood method (ML), and
- the method of probability weighted moments (PWM).

The following theoretical distribution functions were tested for their applicability: E1 – (Gumbel) with parameter estimation (MOM, ML, PWM), GEV – (Generalised extreme value) with parameter estimation (MOM, ML, PWM), ME – (Rossi) with ML parameter estimation, LN3 – (3-parameter Lognormal) with parameter estimation (MOM, ML, PWM), P3 – (Pearson III) with parameter estimation (MOM, ML, PWM), LP3 – (logPearson III)- with parameter estimation (MOM, ML, PWM), and WB3 – (3-parameter Weibull) with parameter estimation (MOM, ML, PWM).

In order to select the most appropriate fitted distributions, a statistical test is recommended. The testing criterion is computed from the relationship:

$$D + n\varpi^2 + (1 - r_p) \qquad (2)$$

where D is the value of the Kolmogorov test, ϖ^2 is the value of the omega squared test, and r_p is the correlation coefficient between the values sorted in a descending order and their distribution quantiles.

The best fit gives the lowest values of ϖ^2, D and the highest values of r_p, by minimising the value in equation (2).

Table 3 presents the selected three best distribution functions according to the statistical test in DVWK (1999), the first of the three is also the one used for further evaluation of runoff changes for different scenarios. The selected distribution has the highest N-year values of maximum annual mean discharges. The number of selected best distributions is rather high, mostly the GEV, LN3 and ME were chosen. The GEV distribution with parameter estimation using the PWM method was the best and most applied one in our study.

Table 3. Selected three best distribution functions according to DVWK (1999) for the scenarios studied

Scenario	Actual land use	Natural land use	Change in forest composition	Grass over forest	Change in impermeable areas	Grass over farmland	Afforestation of critical hydrotopes
Theoretical distribution	GEV	LN3	GEV	GEV	GEV	LP3	GEV
	ME	P3	ME	ME	GEV	LN3	ME
	LP3	WB3	LN3	P3	LN3	ME	LN3
Method of parameter estimation	PWM	PWM	PWM	PWM	PWM	MOM	PWM
	MLM	PWM	MLM	MLM	MLM	MLM	MLM
	MOM	PWM	MOM	MOM	MLM	MLM	MOM

The N-year values of design floods can be seen from the frequency curves of annual maximum daily average discharges in Figure 4 and Table 4, and show the differences in design floods for the scenarios studied. The lowest values were found for the natural land use scenario. The largest raise over the natural land use for the 100-year design flood exists for the scenario Change in impermeable areas, where the maximum daily average discharge is 455 m³.s⁻¹ compared to 96 m³.s⁻¹ for natural land use. Scenario Change in forest composition has the second worst 100-year value of 411 m³.s⁻¹. Slightly better are scenarios Grass over forest and Afforestation of critical hydrotopes with 100-year design floods equal to 397 and 390 m³.s⁻¹, respectively. Actual land use scenario yields a value of 359 m³.s⁻¹ followed by the only scenario, besides the natural land use, with a smaller 100-year design flood – Grass over farmland – with a value of 346 m³.s⁻¹ that is still 249 m³.s⁻¹ higher than Natural land use but 109 m³.s⁻¹ lower than the worst scenario.

Table 4. Change in N-yearly values of design floods for the scenarios studied after subtracting the value for natural land use

N [years]	Change in design discharge [m³.s⁻¹]						
	2	5	10	20	25	50	100
Actual land use	55.4	85.1	115.2	155.2	170.4	227.1	262.2
Change in forest composition	58.4	89.1	121.2	163.2	179.4	238.1	314.2
Grass over forest	62.4	93.1	124.2	163.2	178.4	232.1	300.2
Change in impermeable areas	80.4	119.1	155.2	201.2	218.4	281.1	358.2
Grass over farmland	43.4	71.1	99.2	134.2	147.4	193.1	249.2
Afforestation of critical hydrotopes	52.4	81.1	111.2	151.2	166.4	221.1	293.2

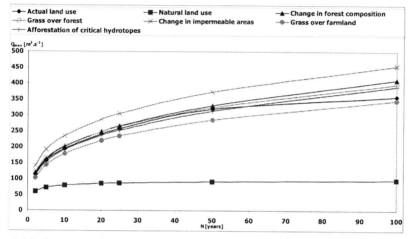

Fig. 4. Maximum annual average daily discharges for actual state and different scenarios

5. Conclusions

The presented results can be used in integrated water resources management for the Hron basin, especially for planning land use and assessing the impact of land use changes on catchment runoff. Change in runoff can be evaluated by comparing average daily, monthly and annual discharges, their statistical characteristics and the spatial distribution of runoff. Runoff components, including overland flow, interflow and baseflow, and the change in water balance components, evapotranspiration and infiltration, can be evaluated separately.

On the other hand, using the results of a distributed rainfall runoff model, one has to keep in mind that the uncertainties of the used approach must be considered. The reliability of the results depends largely on the availability and quality of the input data, on the extent of schematisation of the processes represented by the model, parameterisation of the environment characteristics for the simulated physical processes and on the calibrated global parameters of the model.

Acknowledgement

Authors acknowledge gratefully the VEGA grant agency for support by grants 1/1145/04, 1/2032/05 and 2/5056/25.

References

Beven K (2001) Rainfall - runoff modeling, The Primer. Wiley & Sons, Chichester
Bosch JM, Hewlett JD (1982) A review of catchment experiments to determine the effect of vegetation changes on water yield and evapotranspiration. Journal of Hydrology 55:3-23
Bultot F, Dupriez GL, Gellens D (1990) Simulation of land use changes and impacts on the water balance - a case study for Belgium. Journal of Hydrology 114:327-348
DVWK Regeln 101/1999 Wahl des Bemessungshochwassers. Verlag Paul Parey, Hamburg.
Harr RD (1986) Effects of clear-cut logging on rain-on-snow runoff in western Oregon: new look at old studies. Water Resources Research 22:1095-1100
Hibbert AR (1967) Forest treatment effects on water yield. In: Sopper WE, Lull HW (eds) International Symposium for Hydrology, Pennsylvania. Pergamon, Oxford
Jones JA, Grant GE (1996) Peak flow responses to clear-cutting and roads in small and large basins, western Cascades, Oregon. Water Resources Research 32:959-974
Liu Y, De Smedt F (2004) WetSpa Extension, A GIS-based hydrological model for flood prediction and watershed management. Documentation and user manual, Vrije Universiteit Brussels, Belgium
Mattheusen B, Kirschbaum RL, Goodman IA, O'Donnell GM, Lettenmaier DP (2000) Effects of land cover change on streamflow in the interior Columbia River Basin (USA and Canada) Hydrol. Process. 14:867-885
Niehoff D, Fritsch U, Bronstert A (2002) Land-use impacts on storm-runoff generation: scenarios of land-use change and simulation of hydrological response in a meso-scale catchment in SW-Germany. Journal of Hydrology 267:80-93
O'Callaghan JO (1996) Land use: the interaction of economics, ecology and hydrology. London
Parkin G, O'Donnell G, Ewen J, Bathurst JC, O'Connell PE (1996) Validation of catchment models for predicting land-use and climate change impacts. Case study for a Mediterranean catchment. Journal of Hydrology 175:595-613
Riebsame WE, Meyer WE, Turner BL (1994) Modeling land use and cover as part of global environmental change. Clim. Change 28:45-64
Stednick JD (1996) Monitoring the effects of timber harvest on annual water yield. Journal of Hydrology 176:79-95
Szolgay J, Hlavčová K, Kohnová S, Parajka J, Skalová J, Kubeš R, Zvolenský M, Danihlík R, Papánková Z, Horvát O, Gaál L (2004) Analysis of possible changes in runoff in the Hron River basin due to land use change (Report for the Slovak Ministry of Environment), SUT Bratislava
Veldkamp A, Fresco LO (1996) CLUE: a conceptual model to study the conversion of land use and its effects. Ecol. Modell. 85:253-270
Wang ZM, Batelaan O, De Smedt F (1996) A distributed model for water and energy transfer between soil, plants and atmosphere (WetSpa). Phys. Chem. Earth 21(3):189-193

BUNDLED SOFTWARE FOR LONG-TERM TERRITORIAL FORECASTS OF SPRING FLOODS

EUGENE GOPCHENKO[1], JANNETTA SHAKIRZANOVA
Odessa State Environmental University, Lvovskaya15, 65016 Odessa, Ukraine

Abstract. A method for long-term territorial forecast of spring floods and bundled software for forecasting maximum flood discharges is demonstrated by the example of the Seim River.

Keywords: The long-term territorial forecasting, maximum discharge, runoff depth of spring flood

1. Introduction

The currently used methods for long-term forecasting of the maximum discharge and runoff depth of the spring flood are basically designed for a stream-gauging station with an observed flow series. However, equally relevant, with respect to disastrous floods in particular, is a forecast of the maximum runoff for the region as a whole, including ungauged catchments. Such a problem can be solved by the development of methods for territorial forecast of the spring maximum flood within a region, where similar conditions for the processes of flood formation exist.

The methods of long-term forecast of runoff depth of the spring flood are mostly based on the solution of the water budget equation by way of

---

[1] To whom correspondence should be addressed. Eugene Gopchenko, Odessa State Environmental University, Hydrology Department, Lvovskaya15, 65016 Odessa, Ukraine; e-mail: gidro@ogmi.farlep.odessa.ua

local or territorial dependencies of runoff depth over the catchment area (i.e. volume) on total water-storage of snow cover and spring precipitation, as well as on coefficients of soil moisture and frost zone or determination of the melt water loss by a runoff coefficient.

For the first time the question of the territorial forecast of runoff depth of the spring flood was considered as far back as in the 1950s by Komarov (1955). In recent years a numerical model for long-term forecast of runoff depth of the spring flood, "Sloy-2", was developed Sosedko (Cochelaba et al. 1990) and is currently used in Ukraine.

The development of methods for territorial forecast of the maximum flood discharge encounters certain difficulties, related to both the maximum discharge and its components being affected by the catchment area. The existing methods of forecast of the spring flood maximum discharge are based on the dependency of the maximum discharge (Q_m) on the spring runoff depth (Y_m). The territorial variant of such type of dependencies was for the first time considered by Zmiyova (Zmiyova and Klimova 1983).

2. The proposed method for long-term forecast of spring flood characteristics

The proposed method is based on establishing regional dependencies of the maximum discharge and runoff depth of the spring flood on water storage in the catchment, i.e. the water equivalent of snow cover before the beginning of spring and rainfall during the period of the spring flood, normalised by (k) their average annual values as:

$$k_m = f(k_P) \qquad (1)$$

where k_m is the coefficient of the runoff depth of the spring flood ($k_m = Y_m / Y_0$, where Y_m and Y_0 are the runoff depth of the spring runoff and its annual average, respectively) or the maximum discharge of the spring flood ($k_m = q_m / q_o$, where q_m and q_o are the maximum spring flood specific discharge and its annual average, respectively); and, k_P is the coefficient of total water input to the catchment during the period of the spring flood:

$$k_P = (S_m + P_1 + P_2)/(S_0 + P_{1_0} + P_{2_0}), \qquad (2)$$

where S_m and S_0 are the maximum water storage of snow cover and its annual average, respectively;

P_1 and P_{1_0} are the precipitation during the period of snowmelt and its annual average, respectively; and,

P_2 and P_{2_0} are the precipitation during the period of spring flood recession and its annual average, respectively.

Methods for such territorial forecasts of maximum discharge of the spring flood were presented in a manual of hydrological forecasting (Guidance on hydrological forecasting 1989). However, territorial forecasting can be generalised by using an expected annual flow q_0 which is derived from the actual observed runoff series. Such methods of forecasting can not be applied to ungauged rivers, mainly due to the lack of reliable methods for calculating q_0.

The forecasting dependencies (eq. 1) do not reveal a well defined relationship between the normalised coefficients of runoff depth (of maximum discharge) of spring floods and the total equivalent of water in catchments, which can be explained by the multiple-factor nature of the processes of spring runoff formation. To typify spring floods according to their water content the authors used a multivariate statistical model, and a discriminant function DF, which accounts for complex factors influencing the conditions of spring flood formation. Function DF can be written as:

$$DF = a_0 + a_1 x_1 + a_2 x_{2+...} + a_m x_m, \qquad (3)$$

where $A = (a_0, a_1, ..., a_m)$ is a coefficient vector of the discriminant function; $X = (x_1, x_2, ..., x_m)$ is a measured characteristic (the factor vector); and m is the number of measured characteristics.

In the factor vector of discriminant functions, such characteristics of the flood are entered as the total water storage in the catchment, which contributes to the spring flood formation, the soil moisture and frost zone, expressed by normalised coefficients, as well as meteorological characteristics of the winter and the intensity of spring snowmelt.

The sign of the discriminant equation (either greater or smaller than zero) of rivers in the territory makes it possible to attribute any flood to one of the groups, which conform to certain types of spring floods - wet, average or dry. The established forecast relationships (eq. 1) are described by a polynomial of the n-th degree ($n \leq 3$), i.e.

$$k_m = b_0 + b_1 k_p + b_2 (k_p)^2 + b_3 (k_p)^3, \qquad (4)$$

where b_0, b_1, b_2, b_3 are the coefficients of the polynomial.

For regions with similar conditions of spring flood formation the discriminant equations are steady and the curves can be generalised for rivers in such regions. Besides, as the analysis and checking of forecasts for spring runoff depth and the maximum flood discharge in the basins of rivers in the southern part of Ukraine have proved, coefficients of discriminant equations (3) and coefficients of the polynomial (4), which describe the forecast dependencies, are similar. This can be indicative of the fact that statistical laws for sharing the values of maximum discharge and runoff depth of spring flood are similar.

On the day of forecast issue the unknown coefficients, which are treated as members of the vector of discriminant functions, are estimated by an average value (or by weather forecast), as well as by means of introduction of corresponding adjustments. This pertains to such flood factors as maximum water storage of snow cover, precipitation during the period of snowmelt P_1 and the period of spring flood recession P_2. The authors suggested a method for determining the expected values for spring precipitation P_{1_0} and P_{2_0} according to linear functions $P_1 = k_1 t_r$ and $P_2 = k_2 t_f$, where t_r is the duration of overflow, and t_f is the duration of flood recession. The angular coefficients k_1 and k_2 depend on the latitude of catchments, but the calibrated durations t_r and t_f are expressed integrally from catchment dimensions.

Spatial generalisation of the normal spring runoff depth and average annual values for maximum water storage of snowpack in eq. (1) is carried out on the basis of drawing maps of change of these values across the territory. For determination of the maximum specific spring runoff q_o for ungauged rivers the authors suggested methods for its calculation within the framework of a model of typical single-mode flood hydrograph as:

$$q_0 = \frac{q_0'}{(A+1)^{n_1}} \delta_1 , \qquad (5)$$

where q_0' is an average annual maximum overland inflow ($q_0' = k_0 Y_0$), Y_O is the normalised annual runoff depth for the period of the spring flood, and k_0 is the coefficient of slope transformations or the rush-of-the-flood coefficient (Gopchenko and Gyshlya 1989):

$$k_o = \frac{n+1}{n} \frac{1}{T_o} , \qquad (6)$$

where T_o = duration of the overland flow influx, and $(n+1)/n$ is the coefficient of overland runoff non-uniformity; n_1 is the exponent of reduction of the maximum value across the catchment area A; and δ_1 is the coefficient reflecting the presence of lake storage.

Spatial generalisation of eq. (5) parameters can be carried out by mapping the values of the normal annual spring runoff Y_o and coefficients of slope transformation k_o or duration of influx T_o. The influence of local factors (presence of wetlands and forest cover) on T_o must be eliminated.

The expected normalised coefficient values for the spring runoff depth and maximum discharge of the flood k_m for each year are represented in the form of maps of their changes across the territory. Simultaneously with this map, probabilities (*P p.c.*) for runoff depth of forecast values and the maximum annual spring flood discharge are drawn.

Derivation of the forecast value for the spring flood depth and discharge is carried out as follows:

$$Y_m = k_m Y_o \text{ (mm)} \tag{7}$$

and the maximum spring discharge -

$$Q_m = k_m q_o A \text{ (m}^3\text{/s)} \tag{8}$$

where Y_0 and q_0 – are the normalised runoff depth (mm) and the maximum specific spring flood discharge (m³/s/km²), respectively, which are defined as arithmetic means of annual observations, but if such observations are not available, q_0 can be determined from eq. (5).

When forecasting maximum discharges and spring runoff depths for ungauged rivers, estimation of permissible error for the forecast is also relevant. For example, for some rivers with low relief catchments, relatively simple relationships for the value of permissible error as a function of the catchment area (for maximum discharge) or of the latitude of the geometrical centre of gravity of the catchment (for spring runoff depth) were determined.

The proposed model of long-term territorial forecasts for characteristics of spring floods has been applied to the rivers of the Upper and Middle Dnieper, the Western Dvina, and the Neman, as well as to the rivers in the southern part of Ukraine. The correctness of test forecasts varied within from 86-95 to 65-77%, respectively, and the performance criterion S/σ varied from 0.40 to 0.76 for forecast lead times of 30-50 days.

3. Application of the forecast method

A computer model developed at the Odessa State Environmental University (Gopchenko et al. 2005) produces long-term territorial flood forecasts on the basis of the proposed method, which is suitable for forecasting the maximum discharge of the spring flood in low relief plain rivers and its spatial representation in the form of a map of the expected maximum flood coefficients and their probabilistic estimation in the on-line mode in the offices of the hydrometeorological service.

The database for making long-term forecasts of the maximum spring flood discharge by the proposed computer model includes basic catchment data (physiographic features of the catchment and average annual values of factors influencing flood generation) and on-line hydrometeorological information.

Calculation of the average maximum water storage in snow cover for a catchment, for the date of forecast S_m, is carried out according to the following scheme:

$$S_m = [S(1-a_f) + k_f S a_f] + \overline{\Delta S}, \qquad (9)$$

where S is the maximum water storage in snow cover (determined by field measurements) on the date of forecast, mm; a_f is the forest cover of the catchment, as a fraction; k_f is the coefficient of snow accumulation in forests; and, $\overline{\Delta S}$ is the normal addition to maximum water-storage of snow cover, mm.

The value of $\overline{\Delta S}$ is determined depending on the expected weather conditions (air temperatures determined by meteorological weather forecasts in March or even in February, as a value which is equal to, or greater or smaller than the annual normal) and geographical location of the snow cover water storage measurements

$$\overline{\Delta S} = b_1 + b_2(\varphi - \varphi'), \qquad (10)$$

where φ is the geographical latitude of the point of measurement of snow cover water storage (degrees NL); and, φ' is the relative latitude (degrees NL), to which the values of $\overline{\Delta S}$ are reduced. Coefficients b_1 and b_2 in eq. (10) are defined for any date of forecasting.

In the absence of on-line information on snow surveys at specific points, the data on the snow cover are retrieved from maps of snow cover distribution across the territory.

Information on the depth of frost zone (in the absence of measurements at certain points) is produced for specific dates using the following relationship

$$L_j = L_{j(\varphi')} + b_\varphi(\varphi - \varphi') \qquad (11)$$

where L_j is the frost zone depth at a point J, in cm; $L_{j(\varphi')}$ is the frost zone depth in a provisional latitude φ'; in degrees NL; φ is the latitude of the point of measurement of depths of frost zone, degrees NL; and, b_φ is the tangent of the angle between the flow line and the abscissa axis.

Similar dependencies are used when runoff observations are missing; reconstruction of values for runoff from September of the previous year to January of the current year (in litres/s/km²)

$$(q_{09-01})_i = (q_{09-01})_{i\varphi'} + b_\varphi(\varphi - \varphi'), \qquad (12)$$

where $(q_{09-01})_i$ is the average runoff from September of the previous year to January of the current year in the i-th catchment; $(q_{09-01})_{i\varphi'}$ is the average runoff from September of the previous year to January of the current year at a provisional latitude φ' (degrees NL); φ is the latitude of centroid of the catchment (in degrees NL); and, b_φ is the tangent of the angle between the flow line and the abscissa axis.

The relationships described by eqs. (11) and (12) are produced automatically by the computer program for a given date of forecast.

The operational procedure for running the program (as a case study for the Seim River) is as follows:

- To start the program it is necessary to run the Seim.exe file from the directory where the program is installed. After starting the file, the dialogue box "Seim" is opened, in which the desired year and a date for making the forecast can be chosen;

- the necessary initial data are entered and the reconstruction of missing meteorological and hydrological data is executed;

- forecasting and estimation of the maximum spring flood discharge is executed;

- tables of initial data and forecasted values of the maximum spring flood discharge are printed during their transmission to the Excel environment; and,

- maps of forecasted values of the maximum modular (normalising) coefficients of the spring flood and their expectancy within the annual period are drawn.

- Isolines of expected maximum modular coefficients are constructed automatically, with a specified spacing 0.1, 0.2 or 0.5 (for maps of expectancy, the spacing is 5, 10 or 20%), depending on the range of change of forecast values for maximum modular coefficients and their expectancies across the territory. When producing the chosen series of isolines, it is possible to assign the order of joining the points independently, or, when necessary, to construct several separate isolines of a specified value or split the series into several auxiliary ones. The plotting of the maps is also possible by means of Microsoft Office Excel.

Examples of maps for the anticipated modular coefficients of the maximum spring flood discharges and their expected values for the Seim basin and the spring flood of 2005 are shown in Figs. 1 and 2.

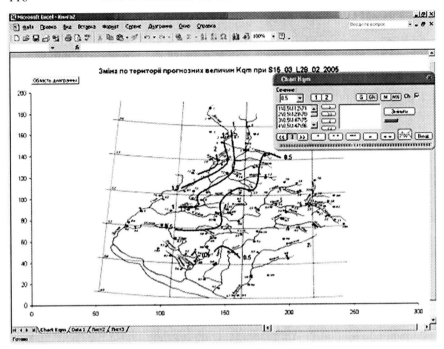

Fig. 1. Variation of forecasted modular coefficients for the maximum spring flood discharge in the studied territory in 2005

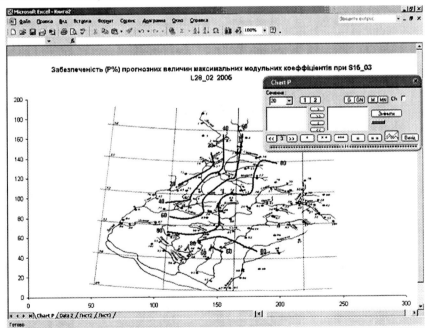

Fig. 2. Variation of expected values of the forecasted modular coefficients for the maximum spring flood discharge in the studied territory in 2005 (%)

The maps offer an immediate description of vast territory as a whole, assessment of zones with potential threat of high spring flood and, what is of particular importance for ungauged rivers, determination of the probability of occurrence of the forthcoming maximum discharge (runoff depth) in the annual series context. The maps allow making a forecast of runoff depth or maximum flood discharge at any point of the territory, regardless of the availability of flow measurements.

4. Conclusions

The suggested method can be recommended as a scientific basis for territorial forecasting of the maximum discharge and runoff depths of spring floods in any region. Further studies should address the use of GIS-technology for creating maps of the anticipated inundated area in the territory during the period of spring flooding.

References

Cochelaba E, Okorsriy V, Sosedko M (1990) Mathematical modelling of processes of forming of spring flow on the territory of Poles'ya taking into account the snowmelt phenomena (in Russian). In: Proceedings of UkrNYGMY, No. 235, pp 3-18

Gopchenko E, Gyshlya A (1989) Hydrology with fundamentals of land-reclamation (in Russian). Gydrometeoizdat, Leningrad, Russia

Gopchenko E, Shakirzanova J, Andreevskaya G (2005) Computer facilities of spatial generalisation of the expected descriptions of maximum spring flow of the plain rivers (using the example of the Desna basin) (in Ukrainian). Meteorology, climatology and hydrology, No. 45, Odessa, Ukraine

Guidance on hydrological forecasting (1989) No. 1, Long-term forecasting of elements of the water mode of rivers and reservoirs (in Russian). Gydrometeoizdat, Leningrad, Russia

Komarov V (1955) Long-term forecasting of spring flow of rivers of the black earth area ETS on the basis of territorial general dependences (in Russian). Gydrometeoizdat, Leningrad, Russia

Zmieva E, Klimova V (1983) Long-term forecasting of maximum spring flow (in Russian). In: Proceedings of Gydromettsentra of USSR, No. 265, pp 3-16

A STOCHASTIC APPROACH TO FLOOD WAVE PROPAGATION ON THE CRISUL ALB RIVER

RADU DROBOT[1], CORNEL ILINCA
Technical University of Civil Engineering, Bucharest, Romania

Abstract. The classic Muskingum model for flood wave propagation is extended to long reaches of receiving tributaries. The flood formation conditions defined by the lag factors are treated distinctly from the river characteristics, reflected by the propagation coefficients. In the case of the Crisul Alb River, three floods (1981, 1995-1996 and 2000) were examined. Due to their different genesis, different lag values were obtained. The propagation coefficients were statistically treated and characterised by the Beta distribution.

Keywords: floods, propagation, lag factors, stochastic coefficients

1. Introduction

For forecasting needs, the Muskingum model is a robust tool offering many possibilities, which are not as yet well known. The aim of this paper is to prove that the Muskingum model may:

- Be applied directly to long reaches;
- Account explicitly for tributaries;

---

[1] To whom correspondence should be addressed. Radu Drobot, Technical University of Civil Engineering Bucharest, Bd. Lacul Tei 122-124, Sector 2, RO-020396, Romania; e-mail: drobot@utcb.ro

- Use a statistical distribution for the coefficients of the flood wave equation; and,
- Define the maximum downstream flood wave corresponding to a given cumulative probability.

The classic Muskingum model expresses the downstream discharge q_i at the moment i as a convex linear combination of the upstream discharges Q_i and Q_{i-1} at the current and previous time steps and the downstream discharge q_{i-1} at the previous moment:

$$q_i = c_0 Q_i + c_1 Q_{i-1} + c_2 q_{i-1} \tag{1}$$

The coefficients c_0, c_1 and c_2 are positive $c_0 \geq 0$, $c_1 \geq 0$, $c_2 \geq 0$ and their sum, like in any convex linear combination, equals one:

$$c_0 + c_1 + c_2 = 1 \tag{2}$$

The coefficients c_0, c_1 and c_2 depend of the propagation time K of the considered reach, the time step Δt and the attenuation coefficient X. In order to ensure the computation stability the parameter X and the ratio $\Delta t / K$ must meet some constraints, of which equivalent is a polygon of the admissible solutions (Musy 1998).

In many practical situations, the distance between two successive hydrometric stations, where accurate hydrographs are available, is quite large. For a good approximation of the rising limb of the hydrograph a small time step is necessary. As a result, the ratio $\Delta t / K$ is small, with the parameters being outside the admissible domain for the usual values of X.

The most common solution in this case is to divide the river reach into a number of sub-reaches, resulting in a larger ratio $\Delta t / K$ for each sub-reach and returning into the domain of the admissible solutions. Still, at the end of each sub-reach the real hydrograph is not known; in these conditions, the parameters cannot be validated for every sub-reach, having thus average values of the parameters for the whole reach considered.

Instead of repeating flood propagation calculations in each sub-sector, another solution can be obtained by computing directly the downstream hydrograph as a function of the upstream hydrograph, but taking into account modifications of the hydrograph along the reach.

2. Presentation of the mathematical model

In the initial analysis, two sub-reaches for the flood wave propagation will be considered.

Eq. (1) is treated as a linear correlation, with the coefficients c_0, c_1 and c_2 representing the weights of the discharges Q_i, Q_{i-1} and q_{i-1}. The downstream discharge q_i^1, at the end of the first sub-reach, can be expressed from eq. (1) as:

$$q_i^1 = c_0^1 Q_i + c_1^1 Q_{i-1} + c_2^1 q_{i-1}^1 \qquad (3)$$

where c_0^1, c_1^1, c_2^1 are discharge weights for the first sub-reach.

Using the same relation, the downstream discharge q_i^2 can be written as:

$$q_i^2 = c_0^2 q_i^1 + c_1^2 q_{i-1}^1 + c_2^2 q_{i-1}^2 \qquad (4)$$

or taking into account eq.(3):

$$q_i^2 = c_0^2 \cdot c_0^1 \cdot Q_i + \left(c_0^2 \cdot c_1^1 + c_0^2 \cdot c_2^1 \cdot c_0^1 + c_2^1 \cdot c_0^1\right) Q_{i-1} + \left(c_0^2 \cdot c_2^1 \cdot c_1^1 + c_2^1 \cdot c_1^1\right) Q_{i-2} +$$

$$+ \left(c_0^2 \cdot c_2^1 \cdot c_2^1 + c_2^1 \cdot c_2^1\right) q_{i-2}^1 + c_2^2 \, q_{i-1}^2 \qquad (5)$$

If one assumes that the propagation time in each sub-reach is almost the same and equal to the time step Δt, the discharge q_{i-2}^1 passing through the downstream cross-section one time step before and its influence can be included in the discharge q_{i-1}^2. Eq. (5) can thus be written in the following form (Drobot and Moldovan 1996):

$$q_i^2 = C_0 Q_i + C_1 Q_{i-1} + C_2 Q_{i-2} + C_3 q_{i-1}^2 \qquad (6)$$

For a river reach divided into 3 sub-reaches, the downstream discharge q_i^3 at the end of the 3rd sub-reach is obtained from a similar relation:

$$q_i^3 = C_0 Q_i + C_1 Q_{i-1} + C_2 Q_{i-2} + C_3 Q_{i-3} + C_4 q_{i-1}^3 \qquad (7)$$

By generalisation, the theoretical discharge q_i^n at the end of the nth sub-reach, or what is the same thing the discharge q_i at the end of the reach, is:

$$q_i = q_i^n = C_0 Q_i + C_1 Q_{i-1} + \ldots + C_n Q_{i-n} + c_{n+1} q_{i-1}^n =$$
$$= \sum_{j=0}^{n} C_j Q_{i-j} + C_{n+1} q_{i-1}^n = \sum_{j=0}^{n} C_j Q_{i-j} + C_{n+1} q_{i-1} \qquad (8)$$

If the number of sub-reaches into which a reach is divided is large, the first coefficients in the relation (8) could be negative. This means that the influence of the current or immediately previous discharge from the upstream end is not transmitted during the current time step to the downstream end; in such cases the corresponding coefficients will equal null. At the same time, in order to ensure the uniqueness of the solution all the other coefficients of the equation, except one, will be also considered null. Thus, eq. (8) will assume the form:

$$q_i = q_i^n = C_j Q_{i-j} + C_{n+1} q_{i-1} \; ; \; j < i \qquad (9)$$

where Q and q are the upstream and downstream discharge, respectively, and k is a lag index.

A similar relation is obtained when there are tributaries along the examined reach; additional terms having the general expression

$$\sum_{j_1=k_1}^{n_1} \alpha_{j_1} T_{i-j_1},$$

where T is the tributary discharge, will be introduced.

Thus, for one tributary only, eq. (9) becomes:

$$q_i = q_i^n = C_j Q_{i-j} + \alpha_{j_1} T_{i-j_1} + C_{n+1} q_{i-1} \qquad (10)$$

By generalisation, for m tributaries the previous equation can be written as (Drobot 2003):

$$q_i = q_i^n = C_j Q_{i-j} + \sum_{r=1}^{m} \alpha_{j_r}^r T_{i-j_r}^r + C_{n+1} q_{i-1} \qquad (11)$$

3. Calibration of the mathematical model

Equation (10) as well as eq. (11) can be interpreted as a multiple linear correlation, of which coefficients are obtained using the least square method. In the case of one reach, the objective function is:

$$(min) Z_1 = \sum_{i=1}^{N} \left(q_i^c - q_i^m\right)^2 = \sum_{i=1}^{N} \left(c_0 Q_i + c_1 Q_{i-1} + c_2 q_{i-1}^c - q_i^m\right)^2 \tag{12}$$

where N is the number of measured discharges.

For the optimal configuration of the parameters, the objective function is minimised, meaning that:

$$q_i^c \approx q_i^m \tag{13}$$

If this statement is true for index i, it can be considered true also for the moment $i-1$, and, consequently, one can write:

$$q_{i-1}^c \approx q_{i-1}^m \tag{14}$$

This approximation is convenient in the objective function (12) in order to avoid recursive formulations of the downstream discharge (Drobot 1987). As a result, by replacing q_{i-1}^c in the objective function with q_{i-1}^m, eq. (12) becomes:

$$(min) Z_1 = \sum_{i=1}^{N} \left(c_0 Q_i + c_1 Q_{i-1} + c_2 q_{i-1}^m - q_i^m\right)^2 \tag{15}$$

The constraint (2) implies the perfect equality of the input and output volumes in the reach or the equality of the sums of the input and output discharges. In many cases, the output discharges are not considered until the time when they are equal to the input discharges in the reach; in other cases, the water losses in the reach represent a few percent of the total volume. As a result, the sum of the coefficients is close to one but slightly smaller. At the other extreme, if the discharges that are not measured and originate from the rest of the basin are important, the sum of the coefficients is greater than one. In these conditions, the constraint (2) can be relaxed and be formulated as:

$$c_0 + c_1 + c_2 \cong 1 \qquad (16)$$

In fact, this constraint will be totally removed, to provide a better flexibility of the model. Still, the coefficients in eq. (16) must not be negative: $c_0 \geq 0, c_1 \geq 0, c_2 \geq 0$.

A formulation similar to eq. (15) can be obtained in the case of a reach which is divided into n sub-reaches:

$$(\min) Z_1 = \sum_{i=1}^{N} \left(C_j Q_{i-j} + C_{n+1} q_{i-1} - q_i^m \right)^2 \qquad (17)$$

Finally, if the river receives tributaries along the examined reach the objective function is:

$$(\min) Z_1 = \sum_{i=1}^{N} \left(C_j Q_{i-j} + \sum_{r=1}^{m} \alpha_{j_r}^r T_{i-j_r}^r + C_{n+1} q_{i-1} - q_i^m \right)^2 \qquad (17')$$

to which the following constraints are added:

$$C_j \geq 0; \quad \alpha_{j_r}^r \geq 0; \quad C_{n+1} \geq 0 \qquad (18)$$

4. Statistical approach to the flood wave propagation

If only one flood is available for calibration, the coefficients of the flood wave equation are deterministic variables, and they will be used as unique values in the propagation process for future forecasts.

The situation is different if more floods are available. In this case, a part of the flood set is usually used for calibration, while the remaining floods are used for model validation. In fact, in this approach one obtains average values for each parameter. Another approach is to obtain propagation coefficients for each flood; then, all values for a given parameter are statistically processed, which results in a distribution function for each propagation coefficient.

The approach based on statistical distributions is justified by the fact that floods occur in different conditions of the riverbed, like the presence or absence of vegetation, winter conditions, ice jams, concentration of sediments, etc. At the same time, the downstream flood is distinctly influenced by the river characteristics (reflected by the values of the propagation coefficients), and by the specific flood conditions (rainstorm movement direction, rain coverage of the river basin – full or partial, succession of the rain occurrence in different sub-basins) characterised by the lag factors.

It was assumed that each propagation coefficient should be in a limited interval; thus, the Beta distribution was proposed to characterise the coefficient statistics:

$$f(x,s1,s2) = \frac{\Gamma(s1+s2)}{\Gamma(s1)\cdot\Gamma(s2)} \cdot x^{s1-1} \cdot (1-x)^{s2-1} \quad \text{for } 0<x<1 \quad (19)$$

where $s1$ and $s2$ are the distribution parameters.

Having the statistical distributions of the coefficients, two options are possible for obtaining a statistical characterisation of the downstream flood:

- A random set of coefficients is generated, for each entry the downstream flood is calculated, and the maximum discharges of calculated floods are then statistically processed. Other elements, like the flood volume, the duration of the flood rising limb or the total duration of the flood can also be statistically treated.

- One uses the statistical values of each propagation coefficient corresponding to high values of the cumulative probability function (like 90% or even larger values). The use of extreme values instead of average values for the propagation coefficients will lead to the worst case scenario of the downstream flood.

5. The Crisul Alb River case study

The propagation equation for the Crisul Alb River between Gurahont and Chisineu Cris has the following general expression:

$$q_i^{CC} = a\cdot q_{i-1}^{CC} + b\cdot Q_{i-k1}^{G} + c\cdot Q_{i-k2}^{S} + d\cdot Q_{i-k3}^{H} \quad (20)$$

where q_i^{CC} and q_{i-1}^{CC} represent the downstream discharges at Chisineu Cris at time steps i, and $i-1$, respectively;

Q_{i-k1}^{G} is the discharge at Gurahont on the Crisul Alb River, with a lag time $k1$

Q_{i-k2}^{S} is the discharge of the Sebis River with a lag time $k2$, and

Q_{i-k3}^{H} - the discharge of the Chier River with a lag time $k3$.

The results of flood propagation for different years are presented in Figures 1, 2 and 3.

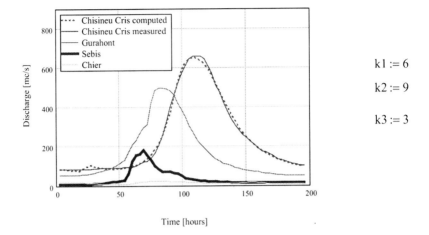

Fig. 1. Flood propagation in the reach Gurahont – the Chisineu Cris (1981 flood)

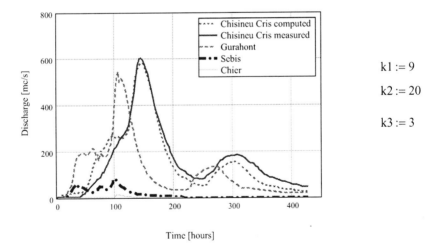

Fig. 2. Flood propagation in the sector Gurahont – the Chisineu Cris (1995-1996 flood)

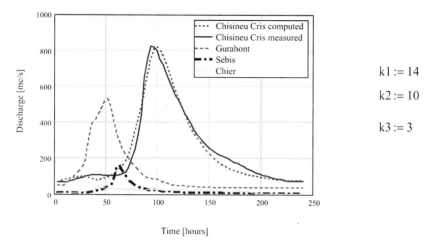

$k1 := 14$

$k2 := 10$

$k3 := 3$

Fig. 3. Flood propagation in the sector Gurahont – the Chisineu Cris (2000 flood)

Using a moving interval $T = \max(k1, k2, k3)$, the propagation coefficients were obtained. Their statistical distributions are presented bellow (Figures 4, 5, 6 and 7).

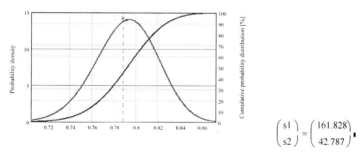

$$\begin{pmatrix} s1 \\ s2 \end{pmatrix} = \begin{pmatrix} 161.828 \\ 42.787 \end{pmatrix},$$

Fig. 4. The statistical distribution of "a" coefficient

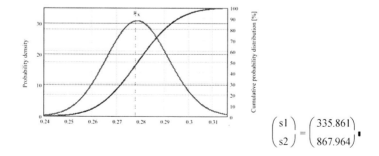

$$\begin{pmatrix} s1 \\ s2 \end{pmatrix} = \begin{pmatrix} 335.861 \\ 867.964 \end{pmatrix},$$

Fig. 5. The statistical distribution of "b" coefficient

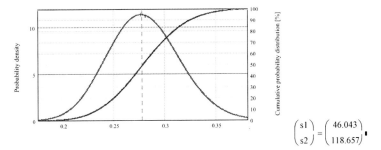

Fig. 6. The statistical distribution of "*c*" coefficient

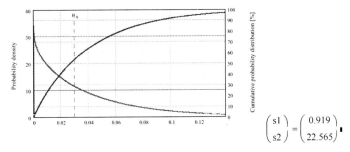

Fig. 7. The statistical distribution of "*d*" coefficient

Based on these distributions, the central parameters (μ_x, m_e, mod) as well as the values of the propagation coefficients corresponding to the probabilities of 90%, 95% and 99% in the cumulative probability distributions were obtained:

$$\mu_x \qquad m_e \qquad \text{mod}$$

$$\begin{pmatrix} a \\ b \\ c \\ d \end{pmatrix} := \begin{pmatrix} 0.789 \\ 0.278 \\ 0.278 \\ 0.030 \end{pmatrix} \qquad \begin{pmatrix} a \\ b \\ c \\ d \end{pmatrix} := \begin{pmatrix} 0.792 \\ 0.279 \\ 0.277 \\ 0.027 \end{pmatrix} \qquad \begin{pmatrix} a \\ b \\ c \\ d \end{pmatrix} := \begin{pmatrix} 0.794 \\ 0.279 \\ 0.277 \\ 0 \end{pmatrix}$$

$$90\% \qquad 95\% \qquad 99\%$$

$$\begin{pmatrix} a \\ b \\ c \\ d \end{pmatrix} := \begin{pmatrix} 0.827 \\ 0.296 \\ 0.325 \\ 0.091 \end{pmatrix} \qquad \begin{pmatrix} a \\ b \\ c \\ d \end{pmatrix} := \begin{pmatrix} 0.836 \\ 0.3 \\ 0.339 \\ 0.118 \end{pmatrix} \qquad \begin{pmatrix} a \\ b \\ c \\ d \end{pmatrix} := \begin{pmatrix} 0.853 \\ 0.31 \\ 0.364 \\ 0.178 \end{pmatrix}$$

Using these values, the propagation of the three floods was calculated by keeping their own lag factors. For illustration, in Figures 8 and 9, the 1981 and 1995-1996 the Chisineu Cris flood waves obtained by propagation on the basis of using the average values μ_x for the coefficients are presented.

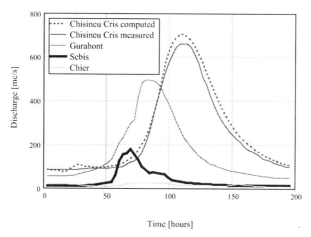

Fig. 8. Propagation of the 1981 flood using the average values μ_x for the coefficients

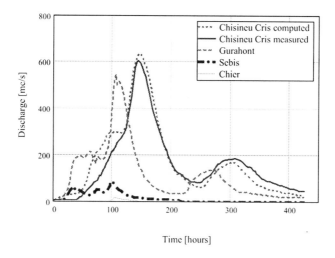

Fig. 9. Propagation of the 1995-1996 flood using the average values μ_x for the coefficients

Finally, in Fig. 10 the downstream flood obtained for the propagation coefficients corresponding to the 99% quantile are shown.

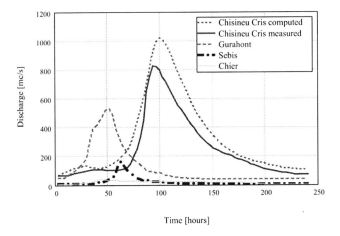

Fig. 10. Propagation of the 2000 flood using 99% quantile values

6. Conclusions

The most difficult problem of flood routing is the evaluation of the lag factors. As a result of calibration for floods with different genesis, a set of models was obtained. In the case of the Crisul Alb River, 3 different models of flood wave propagation were found. For operational purposes, at the beginning of a flood, all the models (characterised by different lag factors) should be used simultaneously. The propagation coefficients can be described by their central values (in order to anticipate the most probable flood downstream) or their extreme values (to forecast the worst downstream flood). According to the flood evolution in the downstream section, before the occurrence of the maximum discharge the most appropriate model will be chosen.

References

Drobot R (1987) Quelques procédés pour l'évaluation des parameters de la méthode Muskingum. Modelling, Simulation and control. C. AMSE Press 8(1):10-24

Drobot R, Moldovan F (1996) Extension of the Muskingum method for n sub-reaches (in Romanian). Scientific Bulletin of Technical University of Civil Engineering Bucharest, 39:31-38

Drobot R (2003) Muskingum model for long reaches receiving tributaries. VI Inter-Regional Conference on Environment and Water "Envirowater 2003" Land and Water Use Planning and Management. Albacete, Spain

Musy A (1998) Hydrologie Appliquée. Ed. HGA, Bucarest

SIMULATION OF FLOODING DUE TO THE CRISUL ALB DYKE FAILURE DURING THE APRIL 2000 FLOOD

ANA NITU, RODICA MIC[1], ROMEO AMAFTIESEI
*National Institute of Hydrology and Water Management,
Bucharest, Romania*

Abstract. The April 2000 flood caused failure of the dyke situated on the left bank of the Crisul Alb River, near the confluence with the Cigher River by Tipari village. The flooding produced in the area delineated by the Crisul Alb, the Crisul Negru and the Teuz rivers, and the Romanian-Hungarian border, was simulated using a two-dimensional model POTOP. The results obtained by modelling consist of maps with land topography, water surface levels, depths and average velocities during various times of flood runoff, maximum water depths, cross-sections, stage hydrographs at every network junction and discharge hydrographs in selected cross-sections.

Keywords: flood, flooding simulation, hydraulic model, unsteady flow

1. Introduction

The flood which occurred in April 2000 caused the failure of the dyke situated on the Crisul Alb river left bank near the confluence with the Cigher river, in the vicinity of Tipari village (Figure 1). The discharge hydrograph of the flow through the dyke breach was determined from the

---

[1] To whom correspondence should be addressed: Rodica Mic, National Institute of Hydrology and Water Management, 97, Soseaua Bucuresti-Ploiesti, 013686 Bucharest, Romania, Email: rodica.mic@hidro.ro

recorded levels during the flood and the rating curves at the gauging stations Ineu and Chisineu Cris on the Crisul Alb river and Chier on the Cigher river using an one-dimensional model *UNDA*.

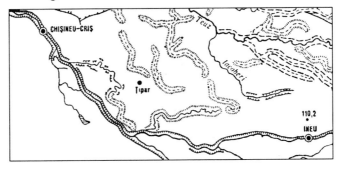

Fig. 1. Crisuri drainage basin – study area

The flooding, which occurred in the area delineated by the Crisul Alb, Crisul Negru and Teuz rivers, and the Romanian-Hungarian border, was simulated using the two-dimensional model POTOP. For modelling the terrain, the model used topographical points and the existing contours of the dykes extracted from 1:25,000 scale maps. The results obtained by modelling consist of maps with elevation contours and water surface, depths and average velocities at various times during the flood, maximum depths, profiles, stage hydrographs at every network node, and discharge hydrographs at chosen cross-sections.

2. Determination of the discharge hydrograph for the dyke breach

The discharge hydrograph from the area between the Crisul Alb and Crisul Negru rivers was determined by means of an one-dimensional hydraulic model *UNDA* (Amaftiesei 1998), which was calibrated on cross-sections of the river channel.

The upstream boundary conditions for the input to this model are two discharge hydrographs on (a) the Crisul Alb river at the gauging station Ineu (the maximum discharge of 690 m^3/s and volume of 155 mil. m^3) and (b) at the junction with the Cigher river (the maximum discharge of 44 m^3/s and volume of 14 mil. m^3). The two-hydrographs were obtained from the measured stage hydrographs and the rating curves for the corresponding gauging stations.

The downstream boundary conditions are represented by the rating curve for the Chisineu Cris gauging station. The elements requiring adjustment are the stage hydrographs from the two considered gauging stations, for which the computations and measurements should be similar.

Because there are no measurements of the flow through the breach of the dyke, two computational scenarios were considered, with and without the dyke failure, respectively. The failure scenario was simulated as a large volume polder with an erodible inflow weir.

The first scenario was considered completed when the calculated stage hydrographs and the measured stage hydrographs, at both gauging stations, reached an acceptable agreement. Obviously, in the Chisineu Cris cross-section, the computed scenario should mimic the observations reflecting the dyke failure (see Figs. 2 and 3).

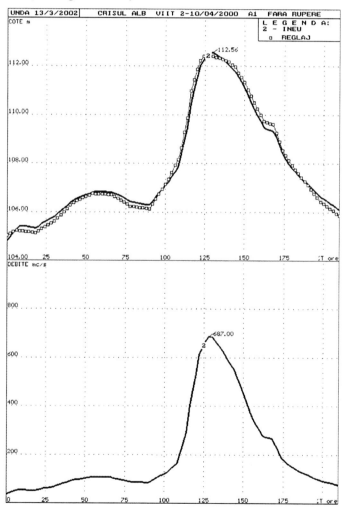

Fig. 2. April 2000 flood: calculated and measured (square symbol) stages (top panel) and discharges at the INEU gauging station (bottom panel)

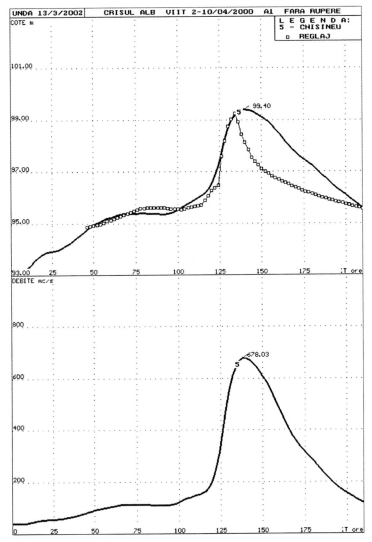

Fig. 3. April 2000 flood: calculated and measured (square symbols) stages (top panel) and discharges at the CHIŞINEU CRIŞ gauging station, without dyke failure (bottom panel)

From this scenario one can reconstruct the maximum discharge at the Chisineu Cris as about 680 m^3/s.

In the scenario that included the dyke failure, the assumed polder parameters (weir crest, length, discharge and hourly rate of erosion) were varied until an acceptable agreement between the measured and calculated hydrographs at the Chisineu Cris cross-section was achieved for the entire hydrograph duration (Fig. 4).

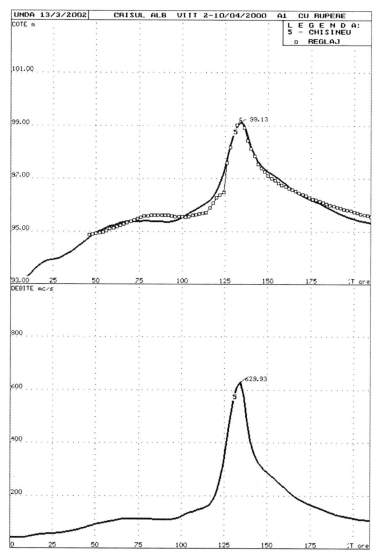

Fig. 4. April 2000 flood: calculated and measured (square symbols) stages (top panel) and discharges at the CHIŞINEU CRIŞ gauging station, with dyke failure (bottom panel)

From this scenario it was possible to extract the discharge hydrographs on the Crisul Alb river, upstream and downstream of the failing dyke section, and derive the discharge hydrograph through the dyke breach. This hydrograph was characterised by the maximum discharge of 400 m^3/s and a volume of 45 mil. m^3 (Fig. 5).

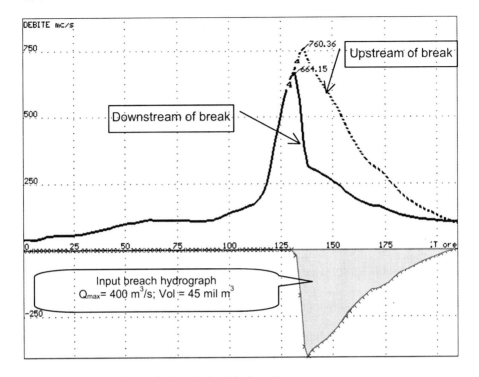

Fig. 5. Discharge hydrograph through the dyke breach

The derived hydrograph through the dyke breach constitutes the upstream boundary condition for the two-dimensional hydraulic model, which is presented below.

3. Application of the POTOP model to simulate flooding caused by the dyke failure

The *POTOP* model (Amaftiesei 2002) serves to simulate floods caused by torrential high-intensity rains in small drainage basins, generally in streams without baseflow. Thus, the main purpose of such simulations is to provide a preliminary delineation of the flash-flood risk area in support of initiating some implementation of flood protection structural measures and issuing flood warning.

The program models unsteady two-dimensional open channel flow by numerical integration of the Saint-Venant equation system comprising the continuity equation and a simplified momentum equation written in Ox and Oy orthogonal directions. Numerical integration is made by linearisation of equations and application of the double sweep solution scheme.

In this study, the *POTOP* model was applied under an assumption that the flood wave produced by the dyke failure can be handled as a two-dimensional problem and the area behind the dyke was initially dry.

The input data file contained: scale factors for the X and Y, Z and T coordinates, network step DX (500 m), time step DT (1 hour), number of time step subdivisions (4), number of time steps (300), number of topographical points used in the interpolation of the field elevations and roughness (3), number of raingauge stations (0), number of downstream boundary condition rating curves (1), number of input hydrographs (1), general roughness (0.1), minimum slope in m/km (0.02), and the number of point marks on maps (8).

The downstream boundary condition for the model was considered for a very low discharge in the Varsand area, where an outlet opening was made in the dyke to drain the inundated area after the flood passed through the Crisul Alb river.

4. Simulation results

The Digital Elevation Model (DEM) of the terrain with values at the network nodes was prepared by the program through linear interpolation in the network of plane triangles formed by the topographical points. In Figure 6, the field map with elevation contours is presented; three following elevations are of particular interest: 100.00 in the vicinity of the dyke breach, 88.50 by the Romania-Hungary border, and 87.50 in the area of coordinate points 34700/65900.

Tipari village was inundated, with an average depth of about 0.5 m, and the maximum depth of 2.5 m (20 hours after the dyke failure). A northern branch of flow by-passed Sintea Mare village and went straight to Adea municipality and reached its southern limit. Also, the northern boundary part of the Chisineu Cris municipality was flooded, with a maximum depth of 0.20 m.

The maximum speed of flood spreading was about 0.2 m/s and the maximum velocities in the flood wave were about 0.4 m/s.

The maximum depths in the studied area during the 10 days after the dyke failure are shown in the map in Figure 7. Knowing that the modelled space has a total surface area of 272 km^2, it could be estimated that at least half of this space was affected by flooding.

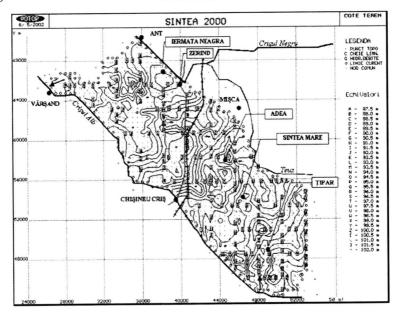

Fig. 6. Terrain elevations in the studied area

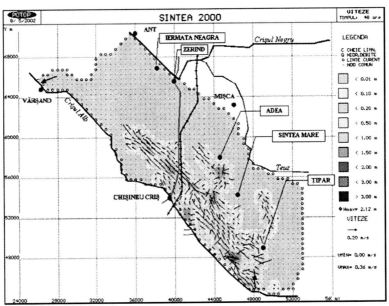

Fig. 7. Depths and water velocities 48 hours after the dyke failure

Stage hydrographs, shown for three network nodes situated at low points of the terrain (Figure 8), highlight the time of flood occurrence, the maximum stage of inundation, and the decreasing or increasing tendency in inundation during various stages of simulation. Thus, in the upstream part,

10 days after the dyke failure, water levels are decreasing (the depth of inundation being about 1 m), while in the area of the border and 10 days after the failure, the water levels are asymptotically increasing.

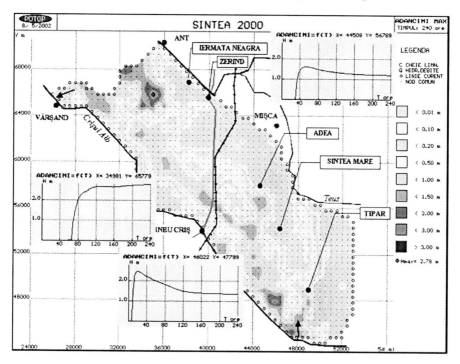

Fig. 8. Maximum depths and stage hydrographs 10 days after the dyke failure

5. Conclusions

In the case of a dyke failure, the reconstruction of hydrographs from gauging stations and the derivation of the dyke breach hydrograph can be done adequately by means of a one-dimensional unsteady flow hydraulic model (*UNDA*), which requires the following input data: (a) stage hydrographs and the rating curves recorded at the neighbouring upstream and downstream gauging stations (including those on the tributaries), and (b) the river bed morphology (cross-sections) between the stations. The flooding of the area behind the failed dyke (i.e., flow into a dry bed) can be simulated by a two-dimensional hydraulic model (*POTOP*), which requires a modest volume of information and support (input data, hardware and software capacity).

References

Amaftiesei R (1998) Instrucțiuni de utilizare a modelului hidraulic unidimensional UNDA (in Romanian; Instructions for using a one-dimensional hydraulic model UNDA). National Institute of Hydrology and Water Management, Bucharest, Romania

Amaftiesei R (2002) Instrucțiuni de utilizare a modelului hidraulic bidimensional POTOP (in Romanian; Instructions for using a two-dimensional hydraulic model POTOP). National Institute of Hydrology and Water Management, Bucharest, Romania

MATHEMATICAL MODELLING OF FLASH FLOODS IN NATURAL AND URBAN AREAS

MICHAL SZYDLOWSKI[1]
Faculty of Civil and Environmental Engineering, Gdansk,
Poland 80-952

Abstract. Computations of flash flood propagation in natural river valleys and urban areas are presented. In the mathematical model of free-surface unsteady flow, the shallow water equations are assumed. In order to solve these equations for a rapidly varied flow, a numerical scheme based on the finite volume method was applied. To estimate the mass and momentum fluxes through calculation cell interfaces the Roe scheme is used. The calculated results are examined against the experimental data. The measured variations of water depth at the control points in the flooded area were available from physical modelling. The experiments were carried out in the hydraulic laboratory of ENEL-CESI in Milan (Italy). Generally a good agreement between the measured and calculated results was observed.

Keywords: flash floods; natural and urban areas; mathematical modelling; shallow water equations; finite volume method; numerical simulations; verification

1. Introduction

Extreme, flash floods in natural valleys and urban areas can be caused by natural events, such as heavy rains, or by failures of flood defence structures, including dams, weirs, sluice gates and flood dykes. Although

---

[1] To whom correspondence should be addressed. Michal Szydlowski, University of Technology, Faculty of Civil and environmental Engineering, Narutowicza 11/12, Gdansk 80-952, Poland; e-mail: mszyd@pg.gda.pl

floods in urban areas are relatively rare, they are much more devastating than in any other areas. Moreover, they can pose a significant threat to human life. Considering these implications, it is clear that numerical simulations of flash flood events play an essential role in hydraulic studies of flooding. Mathematical modelling of flow in natural and urban areas seems to be the main tool for assessing and reducing the risk of extreme flooding. Simulations of flood events are necessary to better protect the cities and increase public safety, and, for example, they can be important for developing emergency plans. The information about potential floods must include such data as:

- time of the flood wave arrival at some points in the valley or city;
- extreme water levels in the flooded area;
- duration and range of flooding; and,
- water depths and velocities in the flooded zone.

The dam-break flows in natural river valleys and urban areas are both rapidly varied and often supercritical. Additionally, flow in a city area has some specific features, such as high turbulence and interaction with buildings and other structures, leading to significant variations of flow profiles. Moreover, flow discontinuities, such as hydraulic jumps, can occur due to wave reflections from walls, for example. Considering all these features it is clear that rapidly varied flow with discontinuities must be modelled well for both natural and urban areas. Concerning topographies of flooded territories and the complexity of city structures, flow simulations in two dimensions (at least) in a horizontal plane are indispensable.

Methods for modelling flash floods in urban areas are rather limited. However, satisfactory modelling results obtained with dam-break flows in natural valleys using shallow water equations (SWE) (Morris 2000) make this approach a promising candidate for urban flows too.

In spite of the gradually changing flow assumption in SWE derivation, these equations can be also assumed to apply to the rapidly varied flow (Abbott 1979). The most widely used methods to solve SWE are the finite difference method (FDM) and the finite element method (FEM). Unfortunately these methods are of little value for rapidly varied flow, when discontinuities (i.e. hydraulic jumps) are present. Oscillations near the discontinuities usually defeat the calculations. In order to avoid that disadvantage a scheme based on the finite volume method (FVM) can be applied. Besides the handling of discontinuities, the numerical solution of SWE for rapidly varied flow poses other numerical problems such as the proper source terms approximation. In the solution used here some special techniques for abrupt changes in bathymetry and the friction term integration are applied.

Extreme floods, like dam-break events for instance, have usually catastrophic consequences and are insufficiently documented. Because of that the results of numerical simulations of such floods are difficult to verify. The simulations presented here address dam-break flood wave propagation along the Toce river valley in the Italian Alps and flash flow through a modelled town. Both problems were investigated in the hydraulic laboratory of ENEL–CESI in Milan (Italy) and laboratory results can be used to verify the computations. The former was a test case in the framework of EC CADAM (Concerted Action on Dam-Break Modelling; http://www.hrwallingford.co.uk/projects/CADAM) project and the latter was addressed in the EC IMPACT (Investigation of Extreme Flood Processes & Uncertainty; http://www.samui.co.uk/impact-project) project.

2. Governing equations and the methods of solution

The system of SWE in conservative form can be written as (Abbott 1979)

$$\frac{\partial \mathbf{U}}{\partial t} + \frac{\partial \mathbf{E}}{\partial x} + \frac{\partial \mathbf{G}}{\partial y} + \mathbf{S} = 0 \qquad (1)$$

where

$$\mathbf{U} = \begin{pmatrix} h \\ uh \\ vh \end{pmatrix}, \quad \mathbf{S} = \begin{pmatrix} 0 \\ -gh(S_{ox} - S_{fx}) \\ -gh(S_{oy} - S_{fy}) \end{pmatrix} \qquad (2a,b)$$

$$\mathbf{E} = \begin{pmatrix} uh \\ u^2h + 0.5gh^2 \\ uvh \end{pmatrix}, \quad \mathbf{G} = \begin{pmatrix} vh \\ uvh \\ v^2h + 0.5gh^2 \end{pmatrix} \qquad (2c,d)$$

In this system of equations h represents water depth, u and v are the depth–averaged components of velocity in x and y direction, respectively, S_{ox} and S_{oy} denote the bed slope terms, S_{fx} and S_{fy} are the bottom friction terms defined by the Manning formula and g is the acceleration due to gravity. Equation (1) can be rewritten in another form as

$$\frac{\partial \mathbf{U}}{\partial t} + div\mathbf{F} + \mathbf{S} = 0 \qquad (3)$$

where the unit vector **n** is assumed as $\mathbf{n}=(n_x,n_y)^T$, and vector **F** is defined as $\mathbf{Fn}=\mathbf{E}n_x+\mathbf{G}n_y$. In order to integrate the SWE in space using the finite volume method the calculation domain is discretised into a set of triangular cells (Fig. 1).

After integration and substitution of integrals by the corresponding sums, eq. (3) can be rewritten as

$$\frac{\partial \mathbf{U}_i}{\partial t}\Delta A_i + \sum_{r=1}^{3}(\mathbf{F}_r \mathbf{n}_r)\Delta L_r + \sum_{r=1}^{3}\mathbf{S}_r \Delta A_r = 0 \qquad (4)$$

where $\mathbf{F}_r$ is the numerical (computed at r^{th} cell–interface) flux and ΔL_r represents the cell–interface length. $\mathbf{S}_r$ and ΔA_r are the components of source terms and the area of cell i assigned to the rth cell–interface.

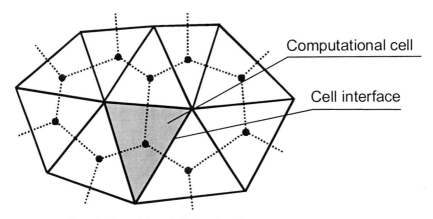

Fig. 1. FVM discretisation of the calculation domain

In order to calculate the fluxes $\mathbf{F}_r$, the Roe (1981) scheme is used. The description of this can be found elsewhere (Glaister 1993; Toro 1997). The source term vector **S** contains two types of terms, which depend on bottom and friction slopes, respectively. Both terms cause numerical integration difficulties. In order to ensure the proper bottom slope approximation this term is up-winded in the same way as fluxes **F** (Bermudez and Vazquez 1994). The second source term (friction) becomes cumbersome when flows of great velocity and small depth are present. In order to avoid the numerical integration problem the splitting technique with respect to physical processes is applied. The numerical algorithm (Szydlowski 2001) is based on a two-step explicit scheme of the finite difference method for integration in time.

MATHEMATICAL MODELLING OF FLASH FLOODS 147

3. Physical modelling

3.1. EXPERIMENTAL INVESTIGATION OF DAM-BREAK FLOW IN NATURAL RIVER VALLEY

A 5-km reach of the Toce River has been reproduced in a 1:100 scale physical model (Testa 1999). The model, built of concrete, is 50m long and 11m wide. It reproduces the riverbed, flood plain and some details of the real geometry such as a reservoir (polder), two bridges, a dam and several buildings.

Control points have been placed along the model to measure the depth variation during experiments (Fig. 2). The model geometry was defined by ENEL-CESI using a Digital Terrain Model (DTM). The physical model was covered by a square grid with interval of 0.05 m. The suggested Manning friction coefficient was n = 0.0162 ($m^{1/3}s$). Initially there was no water in the model. The dam-break was simulated by an extreme flood hydrograph (Fig. 3a), which was generated by a pump system. During the experiment the flow through the inflow section was subcritical. At the outflow section flow became critical after the wave arrival. The rest of the valley margins were high enough to avoid overtopping by water during the flood wave propagation. The spillway of the polder was closed.

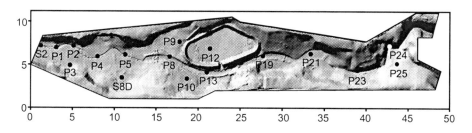

Fig. 2. Geometry of the Toce River model and control point positions (dimensions in m)

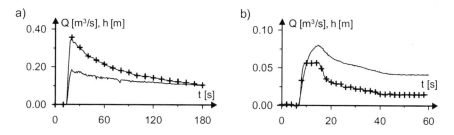

Fig. 3. Discharge (+) and depth (-) measured at the inflow section of a) Toce valley model and b) town model

3.2. EXPERIMENTAL INVESTIGATIONS OF TOWN FLOODING

In order to carry out experiments dealing with urban flooding in the ENEL-CESI hydraulic laboratory, a physical model of the Toce valley was adopted. The modelled town was located in upstream part of the valley, about 5 m downstream of the inflow section. In general, only a 6-m long area (containing model buildings) close to the upstream end of the Toce model was considered in physical and mathematical modelling. In order to simplify the flow structure the 'urban area' is separated from the valley margins by two masonry walls placed parallel to the main model axis. Other model features remained the same as before.

The models of buildings were built as concrete cubes with a 0.15 m side. Two configurations of the building layout were investigated during physical modelling – aligned and staggered. Herein, only the latter is presented. In this test case buildings are placed in a checker board configuration (Fig. 4).

The control points were placed upstream of, among, and downstream of the buildings to measure the depth variation during experiments (Fig. 4). Initially the whole model was dry. The flash flow was simulated as an extreme flood hydrograph (Fig. 3b) at the upstream end of the model. During the experiment the flow through the inlet section was subcritical. At the downstream end of the model town the outflow was subcritical or supercritical depending on the actual flow conditions.

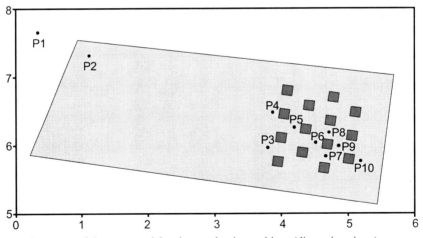

Fig. 4. Geometry of the town model and control point positions (dimensions in m)

4. Numerical computations

4.1. NUMERICAL SIMULATION OF DAM-BREAK FLOW IN NATURAL RIVER VALLEY

In order to simulate the Toce dam-break experiment, the calculation domain is covered by an unstructured triangular mesh composed of 12,965 computational cells (Fig. 5).

Initially the whole domain is covered by a thin water film (thickness of 0.1 mm) simulating dry bottom of the valley. The boundary conditions are imposed in accordance with the experiment. At the inflow section the velocities in the normal direction to the boundary plane are imposed. At the outflow, a boundary condition is imposed by assuming critical flow. The rest of the boundaries are treated as walls (solid boundaries). The buildings were reproduced by modifying bottom levels. The bridges and the second dam were not taken into account during simulations. The calculation is carried out with the time step $\Delta t = 0.01$ s ensuring the computational stability. The simulation time was equal to 180 s.

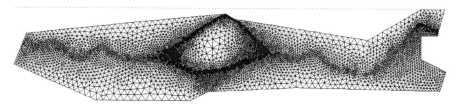

Fig. 5. Numerical mesh for Toce dam-break simulations

The computed and measured depth variations at some control points are shown in Figure 6. Generally the agreement between measured and calculated results is quite good. However, in the upstream part of the valley the water depth is underestimated (e.g., at point P1, in Fig. 2). It can be caused by the boundary condition imposed at the inflow section, where only the velocities are known. The water depth was calculated using the Riemann invariants. The disagreement of results at point S8D (among the buildings) can be caused by the poor approximation of the village located on the right valley margin. However, consistent results were obtained for the next (downstream) points (e.g., P5 in Fig. 2) located upstream the polder. The good agreement for these points located in the middle part of the model was observed despite neglecting the bridge, which was present in the physical model. The depth variation was properly represented again at the points in the downstream part of the model (e.g., P21 and P24, Fig. 2).

This good agreement was observed even though the second bridge and the dam were not accounted for in the simulation and despite of poor approximation of buildings.

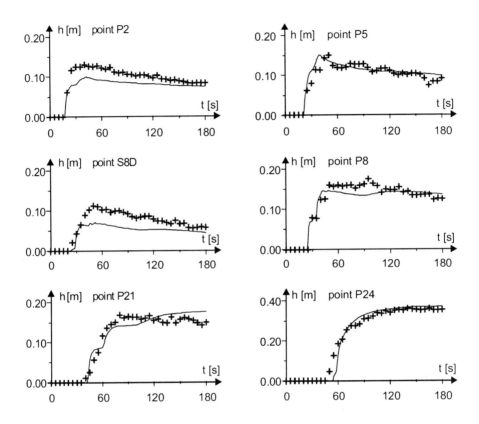

Fig. 6. Calculated (-) and measured (+) depth variations for Toce dam-break simulation

It should be emphasised that the observed and simulated flood waves arrived at the control points at the same time, which proves a proper reproduction of the wave front velocity.

4.2. NUMERICAL SIMULATION OF URBAN FLOODING

In order to simulate the urban flooding experiment the calculation domain was covered by an unstructured triangular mesh composed of 5,185 computational cells (Fig. 7).

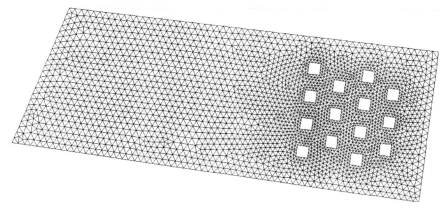

Fig. 7. Numerical mesh for urban flooding simulations

Initially the whole domain was covered by a thin water film (0.1 mm) simulating dry bottom of the flow area. The boundary conditions were imposed according to the experiments. At the inflow cross-section the velocities in the normal direction to the boundary plane were imposed. At the outflow boundary a free outflow condition was imposed. The calculation was carried out with the time step $\Delta t = 0.01$ s ensuring the computational stability. The simulation time was 60 s.

The computed and measured depth variations are shown in Figure 8. In general, the results agreement is very good at all points. Regardless of the control point position the measured depth was reproduced accurately. Point P4 (Fig. 4) was placed upstream of the first row of buildings. As well depth variation and wave arrival time were simulated properly at this location. The flash wave reflected from the buildings front row, resulting in formation of an upstream moving hydraulic jump. It can be seen that the first peak in water levels has not occurred in the computed results. Points P5, P6, P8 and P9 (Fig. 4) were located among buildings. Numerical model gave there a precise representation of water level evolution irrespective of the point location. Point P10 (Fig. 4) lies in the downstream part of the model in a wake zone. At this point some difference between the calculated and measured results can be observed. This point is located between the town model outflow and the downstream boundary, where the boundary condition is unknown. The flow varies here from subcritical to supercritical, depending on the actual hydraulic conditions. Finally, it is not possible to define the influence of the downstream boundary condition on point P10 which may result in a slight disagreement between calculated and measured results.

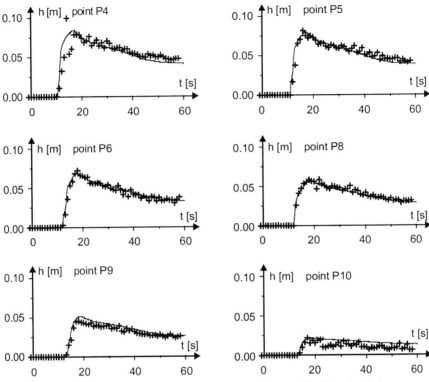

Fig. 8. Calculated (-) and measured (+) depth variation for town flooding simulation

5. Conclusions

The dam-break flood event in the Toce river valley and town flash flooding were simulated using SWE. The computations were verified using experimental data from physical modelling. The agreement between the calculated and measured results was quite good. It can be concluded that although SWE do not truly mathematically represent the free-surface water flow, the numerical model based on these equations is capable of simulating the rapidly-varied flow in both natural and urban areas. Of course, such a model cannot reproduce details of local flow effects, but the main features of flash flow, like reflections from buildings, transitions from supercritical to subcritical flow by means of hydraulic jumps, steep fronts with significant gradients of depth and velocity, and irregular topography or circulation zones can be successfully simulated. The model used for flash flow simulation can be useful in preparation of the emergency and flood protection plans, analysis and mitigation of the flood risk, and design of flood warning systems.

Acknowledgment

The author would like to thank the research group of PIS (Polo Idraulico et Strutturale), ENEL-CESI, Milan, Italy (participant in EC CADAM and IMPACT projects) for experimental data required in this paper. Dr. Guido Testa of CESI SPA is gratefully acknowledged for providing information and allowing to present the CESI results during the NATO Advanced Research Workshop.

References

Abbott MB (1979) Computational hydraulics: elements of the theory of free-surface flows. Pitman, London

Bermudez A, Vazquez ME (1994) Upwind methods for hyperbolic conservation laws with source terms. Computers and Fluids 23:1049-1071

Glaister P (1993) Flux difference splitting for open-channel flows. Int. J. Numer. Meth. in Fluids 16:629-654

Morris MW (ed) (2000) Final Report – Concerted action on dambreak modelling, HR Wallingford Ltd., Wallingford

Roe PL (1981) Approximate Riemann solvers, parameters vectors and difference schemes. J. Comp. Phys. 43:357-372

Szydlowski M (2001) Two-dimensional shallow water model for rapidly and gradually varied flow. Arch. of Hydro-Eng. and Env. Mech. 48(1):35-61

Testa G (ed) (1999) Proc. CADAM Project Meeting, Milan, Italy

Toro EF (1997) Riemann solvers and numerical methods for fluid dynamics, Springer–Verlag, Berlin

FLOOD MODELLING CONCEPT AND REALITY - AUGUST 2002 FLOOD IN THE CZECH REPUBLIC

P. SKLENÁŘ, E. ZEMAN[1], J. ŠPATKA, P. TACHECÍ
DHI Hydroinform a.s, Prague, Czech Republic

Abstract. In August 2002 a disastrous flood occurred in Bohemia (The Czech Republic) and heavily hit all main rivers and cities located on those rivers. The return period of the August 2002 flood exceeded 800 years in some places, and in most cities it was the highest observed flood ever. This flood also affected the Czech capital and the largest city, Prague. Thanks to the hydroinformatics tools applied during the flood situation, valuable information and knowledge regarding flood propagation and flooding extent was available in advance. This paper deals with applications of detailed mathematical models focused on determination of flow characteristics (water level, depth and velocity distribution) during statistically generated floods as well as during actual flood events. These results are used in the next step for delineation of flood zones and also for analysis of flood protection measures needed in the whole river basin. This helps solve problems of continuing urbanisation and economic growth in the areas exposed to floods, associated with further increase of population and the related danger of potential economic and cultural damages. It is necessary to designate these areas as flood plains or areas exposed to special floods, and to regulate adequately their land use. This task is impossible to fulfil without strong hydroinformatics tools, such as the Digital Terrain Models (DTM), hydrodynamic models and GIS tools. The paper concentrates on different approaches (1D and 2D) and different applications of mathematical models as supporting tools for decision

---

[1] To whom correspondence should be addressed. Evzen Zeman, DHI Hydroinform a.s., Na Vrsich 1490/5, 100 00 Praha 10, Czech Republic, e.zeman@dhi.cz

makers in the field, and for assessing flood risks and flood protection in the Czech Republic, verified by an actual catastrophic flood situation.

Keywords: Hydrodynamic Modelling, Flood Risk Analysis, Flood Mapping, Historical floods

1. Introduction – the general situation

For the Czech Republic, floods represent the highest direct risk stemming from natural disasters and they can cause serious crises, which can be associated with high property damages, ecological and cultural damages, and also losses of human lives. The threats of floods are forgotten when one or more generations have missed the actual experience. The urban development in such cases more and more ignores the wisdom and expands activities in the flood planes. The above facts were fully proven during the catastrophic floods which occurred in Central Europe in summers of 1997, 1998, but above all in 2002.

At the beginning of the 1990s, systematic preventive measures were not practiced because the last floods with fatal consequences occurred in the domain of interest at the end of the 19th century. For example, in Prague, there have not been any severe floods observed since 1890 (Q_{100} flood) until the disastrous flood in August 2002 (See Fig.1). Absence of any systematic concept of flood protection and the lack of executed flood protection measures was unfortunately notorious for the period of the last four decades of the last century. The market driven economy also did not recognise the priority of flood protection in the late 1990s, so it was possible to note a low level of awareness of the flood risks and certain stagnation of activities focused on the development of preventive flood measures systems during the last decades. In 1997, when a severe flood struck large areas of the Morava River basin, things have changed. Based on the detailed assessment of this catastrophic flood and experience gained from rehabilitation activities, the Czech Government has assigned a task to prepare a Flood Protection Strategy as a basis for a systematic approach in this area and a basic document for preparation of necessary measures. The Flood Protection Strategy has been issued as a document integrating basic technical knowledge and skills in the region of Central and Eastern Europe with respect to flood management, experiences from previous floods, and took into account technical standards, legislation and organisational framework in formulating further actions to achieve reduction of devastating flood impacts. Another objective was to form a basis for the state or local administration in making decisions concerning the selection of specific flood protection measures and to influence regional development.

The Flood Protection Strategy helped speed up many flood protection measures and improved non-structural measures and flood rescue plans. Unfortunately only a few of the planned flood measures were executed and finished before August 2002.

Since the mid 1990s, mathematical modelling has commenced to be used as a tool for gaining important information, essential for the flood protection planning. Using the example of the Prague Flood Model development we would like to show, how the approach to flood management has been progressing over the years, in the domain of interest, the Czech Republic.

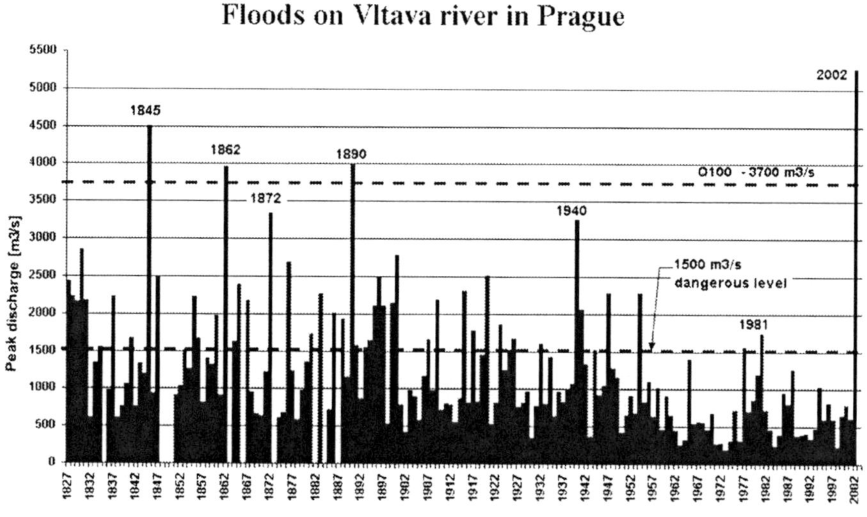

Fig. 1. Floods recorded in Prague during the last 175 years (source CHMI 2002)

2. Flood of August 2002 – description of the historical event

The meteorological situation, which caused the flood, could be described as a rare and unique situation, in which two rather well-developed frontal borders remained for an unusually long time above the domain territory. During the first heavy precipitation period (6.8.-7.8.2002) mainly Southern Bohemia was affected by rainfall with average totals from 130 to 220 mm during two days (the return period from 50 to 100 years). Natural retention of land cover in rural areas of the South Bohemian river basins (the basin of the Vltava River) was exhausted during the first wave of rainfall, causing the increase of discharges to values between Q_{20} and Q_{100} (N return period) in the upper reaches of the Vltava River basin.

During the second wave of precipitation (12.8.-14.8.2002) average totals were about 160 mm, in mountain areas even over 400 mm, but nearly the whole area of Bohemia and a part of Southern Moravia were hit (see Fig. 2). The situation can be characterised as a combination of long duration regional rainfall of low intensity (the return period about 50 years) with extremely intense local rainfall. Characteristics of the second wave of rainfall (long lasting, low intensity and large extent) are typical for some of the previous largest summer floods in the domain, as shown in Fig.1 (1997, 1954, 1890, 1888).

After the first wave of rainfall, the retention capacity of the upper Vltava basin was completely exhausted. The retention capacities of the reservoirs of the Vltava cascade were kept according to the valid operating curves, but the dispatchers were helpless against the volume of water rushing down the basin. However, thanks to the available limited retention volume in the Vltava cascade the dispatchers of the Vltava River Basin Authority were able to keep the outflow of the regulated cascade at a reasonable level of discharge for a while, which enabled Prague authorities to anchor all boats in the inland ports and to install the moveable flood protection barrier in the central area of Prague. This fact also assisted in the way that the other flood protection measures were executed and the evacuation in Prague could take place without chaos. However, due to high volumes coming from all rivers into the cascade the outflow from the cascade became soon uncontrollable. The discharge consisting of two contributing rivers the Vltava and Berounka caused a never recorded flood even in the lower parts of the Vltava river reach. Despite the fact that the main rainfall wave covered nearly the whole effective runoff domain of the Vltava River basin, floods occurred also in the Elbe River basin, mainly in the lower reach downstream of the confluence with the Vltava River. The 2002 flood was assessed as the highest recorded flood in many places (e.g., in Prague since 1827). As an example it can be mentioned that the water level was approximately about 0.7-3.0 m higher than during the previous historically largest flood in Prague (1890) with a considerably lower discharge. Water marks representing the peak flow were set in more than one thousand of places in the river basins exposed to the flood, to serve hydrologic and hydraulic evaluations for updating basic data, formulas and models for further computation, planning and flood management.

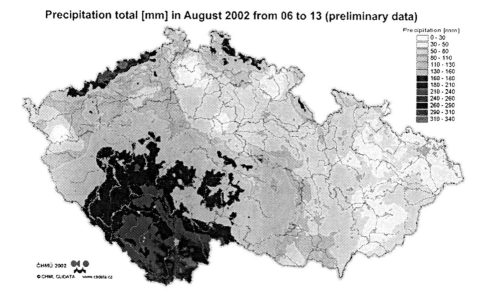

Fig. 2. Precipitation totals, the second wave (source: CHMI 2002)

Unfortunately the hydrologic monitoring network was insufficiently funded in the last century. During the flood most of the river gauge stations were destroyed, and could not transmit or even record data during the unexpectedly high water levels. Although a meteorological forecast was available and quite accurate, hydrological forecast was not fully available in the later stages, because most of the models were out of range and on-line data were hardly available later during the event. Rating curves were not accurate enough or did not exist for high discharges on most of the rivers affected. All the simulations and measures until 2002 were carried out for the highest flood rated as Q_{100}. The flood 2002, rated as approximately Q_{500}, has changed the attitude of all the authorities involved.

3. Modelling tools available in Prague

3.1. AT THE STARTING POINT

In the mid 1990s, the City Council of Prague and the Vltava River Basin Authority decided to improve the preparedness of Prague against floods after the enormous floods occurring in Europe in the mid 1990s. During the period of preparation of the general flood protection scheme for Prague, several alternatives were considered including "do nothing". Among the suggested solutions, one was to build a large scale physical model, which would cost several tens of millions of Czech Crowns. After some deliberations, the mathematical modelling was finally selected for an assessment of the flood protection strategy and the development of measures for the City of Prague and its immediate vicinity. Prague is situated at the confluence of two major rivers in Central Bohemia, the Vltava and Berounka Rivers, with several minor contributing streams in the area. These streams complicate the situation during floods, because backwater effects could threaten some densely populated areas.

The first comprehensive study, finished in 1997, had to be based on a rather complicated hydrological data set. Even though there were unique historical data available, there were also many changes in urbanised areas in the basins as well as anthropogenic changes of the river beds themselves. The study included a hydrodynamic 1D+ mathematical model of Prague (looped 1D model), and also a phase of rainfall-runoff modelling. The hydrodynamic model used the data set from the 1890 flood, and was adjusted for the current status of all parameters in all the included basins, with special focus on the impact of the Vltava cascade. The modelling effort was among others focused on the degree of ability of the cascade to protect the city from floods. The reality is that the volume of all the Vltava cascade reservoirs represents only a fraction of the total volume of the 1890 flood wave. This means that the cascade is able to eliminate the impact of minor floods, but has only slight influence on the catastrophic flood events. For the flood of 1890, which is close to Q_{100} in Prague, the peak discharge reduction is only 10% and it can delay the wave for several hours only.

The 1890 flood data were used at the very beginning of modelling for protecting Prague, at a reasonable frequency level, instead of commonly used theoretical statistically generated flood waves, and it was the first time that the design flood was selected on the basis of an actual historical event. The peak discharge for the Q_{1890} corresponds to approximately Q_{100} in Prague.

3.2. 1D PHASE – THE COMPREHENSIVE STUDY 1994-1997

The 1D hydrodynamic model covered a 33-km river reach, limited by the official borders of the City of Prague, with the upstream cross-section in Vrané nad Vltavou (chainage 70.000 km) and the lower one in Klecany (chainage 37.000 km).

The basic step, making any mathematical model reliable, is calibration, using the recorded event data. The opportunity to use a historical event for calibration was limited: The team of modellers selected for calibration the 1890 flood with an estimated peak discharge of $Q_{max} = 3{,}975$ m^3/s. But, as mentioned earlier, many topographical and urbanisation changes have occurred in the flood plains since 1890. This had to be definitely considered. The dam cascade on the Vltava River changed the runoff pattern by forcing floods through the system of dams with certain operational rules. The rainfall-runoff characteristics are also influenced by both natural and man-made changes. Despite all drawbacks of the 1890 historical event, the team decided to use it for a trend-calibration, because this event was well documented not only in Prague but also on the streams in upper reaches.

A conceptual MIKE-NAM (rainfall-runoff) model has been applied in 6 sub-basins of the main Vltava River tributaries (the Sázava, Malše, Otava, Lužnice, Berounka and upper Vltava) upstream of Prague. The Berounka River is known by steep falling limbs of hydrographs, which when arriving at the confluence with the Vltava River, may cause catastrophic flood events by interference. The rainfall-runoff models were calibrated using selected cross-sections where hydrographs were available for a global hydrodynamic model. Since the information about rainfall distribution, intensity, and space and time variability is more easily available, the team decided to utilise this methodology in order to get more realistic dynamic boundary conditions in the form of hydrographs. The calibrated model set-up enables to simulate any boundary conditions based on the known or assumed rainfall intensities at the inflow cross-sections.

The details of the calibration of the second phase were presented by Zeman et al. (1996). The global model set-up of the Vltava river basin above Prague (Fig. 3) was applied in order to obtain transformed flood waves based on operational rules of all large reservoirs. The operational rules were set-up by the Vltava River Water Basin Authority dispatchers, who were present during the simulation sessions, setting the conditions on structures according to actual cases. This approach was the only one accepting the influence of human decisions during the rainfall-runoff process, when hydraulic structures have an essential impact on hydrodynamic characteristics.

The 1D hydrodynamic model was developed in two phases: in the first phase the river channel model was built, and in the second phase the flood plain model was connected to the river channel model.

Both models were based on an open-channel modelling scheme. There were 390 cross-sections describing the river channel in the 33-km river reach. All the important hydraulic structures in the area of interest (such as bridges, weirs, derivation channels, inland ports, islands and locks) were introduced. All the topographical information was derived from technical maps in scale 1:500, which were available in a sufficient quality.

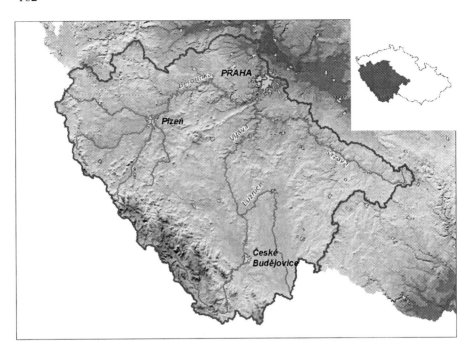

Fig. 3. Catchment of the Vltava River with major tributaries

In the second phase, the flood plain model set-up was derived from the river channel set-up. This model was created on the same river reach as the river channel set-up, but had to cover all areas flooded by extreme floods. The schematisation for the flood plain model set-up was based on MIKE11 1D+ modelling facilities (looped schematisation). This means that the flood plain was divided into particular looped channels, described by cross-sections and covering the corresponding part of the plain. This was the first application of the looped system in the region. The model set-up had to allow the user to model all the most important flood streamflows and storage capacity in order to describe transformation of any flood hydrograph without significant phase and amplitude errors. The methodology adopted by the team seemed to be rather complicated but in mid 1990s, it was the only method which enabled the Vltava Water Basin Authority representatives to evaluate all future design ideas of flood protection measures for Prague. Few parts of the area of interest were studied using 2D models.

The first model generation was utilised by the city council authorities. The flood mapping results were accepted. Comprehensive comparisons with the former flood contours were done and areas, where former and modelled flood lines differed, were identified.

When the comprehensive study was carried out, the available computer technology was not able to support 2D modelling for large areas. The 1D approach should be, in many cases, accurate enough to give answers to some questions. The 1D models are still successfully used in certain types of flood plain areas, and fast computation is still their most advantage. The 2002 flood proved that most of the conclusions based on this study were correct and in some cases provided the crucial point for decision making of the local authorities – the best example are the mobile flood barriers which saved the city centre - Old Town - in August 2002.

3.3. 2D FLOOD MODEL 2000 - 2001

After the successful application of the first 1D Prague Flood Model, the cooperation with the Vltava River Basin Authority and the City Council continued. A newly developed tool for 2D modelling – MIKE 21C- was used to update the Prague Flood Model.

The MIKE 21C package was specifically developed for modelling river morphology. It uses a curvilinear orthogonal grid, which enables to build the mesh, suitable for describing the actual morphology, with a reduced number of computational grid points. Using the 2D approach updates and enhances the information about the water flow in the city during flooding. Maps of flow directions and velocities are obtained for the flooded area, as well as a complete water surface elevation pattern. But the essential reason for using the 2D model is the better accuracy in those parts of the flood plain, where it is not possible to predict exact flow direction, i.e. the reaches, where flow is expected to leave the centreline of the main channel (or side channels) and flow through wide inundation areas. With such flood plain conditions it is extremely work demanding and often impossible to build the 1D model, which would give valid results. There are several such areas in the Prague Flood Model. Two of them, Holešovice-Karlín and Trója, were schematised as separate models in order to test the MIKE 21C modelling package and to obtain some extra information about the flood conditions in these areas. The results revealed some interesting findings, which could not be recognized in the original 1D model – the flow directions often differed from the original estimation, which had the significant influence on overall stream behaviour in these areas. These two mentioned models were later merged, creating the downstream component of a three-component 2D Prague Flood Model 2000. The network contained 1 million computational nodes and consisted of three parts: the lower part Roztoky –Karlín (326,402 nodes), followed by the middle part Centrum (397,736 nodes) covering the central part of the city, and the upstream part called Soutok (357,210 nodes), including the Vltava and Berounka Rivers confluence. The spatial resolution of the grid varied between 2-3 metres; this fine grid was needed for describing the Staré město (Old Town) area with narrow and winding streets, and 5-9 metres in the flood plains, in less important areas. An example of model results is presented in Fig. 4. The 2D modelling technology rapidly increased the usability of models, because the combination of flood depth maps and velocity fields allowed assessing endangered zones during floods and such results are very valuable for preparation of flood management plans and evacuation planning maps.

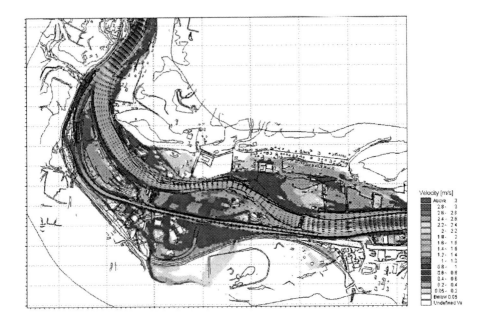

Fig. 4. Example of MIKE 21C output: flow velocities in the area of Císařský Island

Parts of the 2D model were used for assessing various flood protections and city development planning in the flood-prone area, and for the first time they were also used for validation of design of particular flood measures.

A disadvantage of the 2D approach is long computational time for dynamic simulations, which vastly exceeds that of the 1D approach. Nevertheless the benefits of the 2D approach were beyond any doubts and it was obvious, that it is only a question of time, before the hardware development will enable fully dynamic simulations without severe time constraints.

Contingency planners and managers used all the generated maps during the 2002 flood. Moreover, mathematical models presented here were used in an operational mode, providing the technical support for the Flood Protection Council of Prague. Emergency and rescue forces used the flood maps with clear indications of depths, water levels and in many cases also velocity fields simulated for the discharges, which were usually lower than the catastrophic ones in August 2002. But nevertheless the maps gave them some overview so much needed in times of crisis.

After the 2002 flood, the inundation line generated from simulation results obtained before the flood was compared with the actual line (derived from aerial photographs taken during August 2002 – see Fig.5). Both lines are very close, primarily parallel, so the conclusion was that the modelling results were surprisingly accurate.

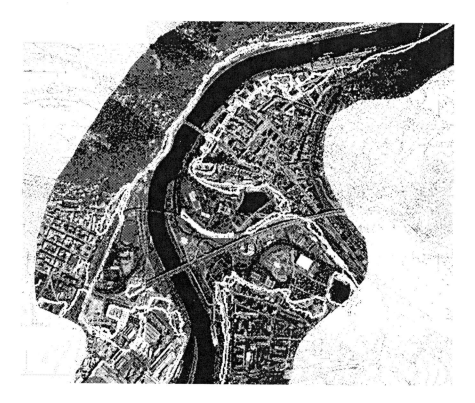

Fig. 5. The comparison of the actual flood extent: grey line = simulated flood line for predicted discharge, white line - the actual flood line 2002/8 as the maximum flooding envelope evaluated from airborne photogrammetry (source: the Elbe River Basin Authority 2002)

3.4. 2D PRAGUE FLOOD MODEL 2003 – UPDATE AFTER THE 8/2002 FLOOD

The flood of 8/2002 was definitely considerably greater than the 1890 flood, used in all the previous flood modelling studies. According to the first estimates, the peak discharge of the 2002 flood reached 5,300 m^3/s, compared to the peak discharge of 4,030 m^3/s of the formerly selected design flood (i.e., a modified flood wave from 1890). As it was impossible to carry out proper calibration of models before 2002 because of lack of relevant water marks and mainly under the changed urban development (as well as river morphology changes in basin), all the calibration factors – hydraulic roughness coefficients - were set at the upper limits of their ranges for safety reasons. The first part of the flood protection scheme, which consist of structural measures was designed with a rather high safety margin. This decision had eventually saved the Staré Město (Old Town) from flooding. But it was really tight, only a few centimetres were left of the original freeboard of 50 cm.

After the flood, the actual calibration of the 2D model was carried out, using the peak discharge that occurred during the 8/2002 flood. A number of reliable flood water marks were established and recorded during the flood, and also the experts from the modelling team had a chance to eye-witness the phases of the flood in Prague.

The evaluation of the 2D model was very detailed and resulted in a general modelling study of the 2002 flood peak using steady state simulation for the peak discharge. The resulting computed water levels were rather accurate; 69% of peaks levels were within 5 cm and over 87% were within 10 cm of the reliable peak records.

3.4.1. *Practical application of the re-calibrated 2D model*

Together with the re-calibration came another request to use the model for evaluation of the Prague Flood Protection Project design works. Prague Flood Protection Project has become a centre of focus. The first basic step of the study was to assess the influence of the completely designed flood protection line (actually the new border reducing the flood-prone area) on flow conditions. Within the framework of this task three scenarios were assessed, varying only by their approach to flood protection of the Central Waste Water Treatment Plant (CWWTP), located on the Císařský Island downstream Prague. The last set of simulations was repeated for the 2002 flood boundary condition taking into account the existing part of flood protection (P0001) and the proposed part (P0002) of the flood protection measures (flood protection of Kampa and Malá Strana).

APPLICATION OF FLOOD MODELLING TOOLS

The hydrodynamic simulations successfully proved that the Prague Flood Protection scheme would not considerably affect the flow conditions during high discharges, because it would cut off mainly the passive parts of the flood plains.

The impact of the Central Waste Water Treatment Plant on water level changes is less than it was assumed (about 10 cm in the case of not overtopping the flood wall); much higher effect is exerted by the roughness variation on the Císařský Island and the whole flood plain area due to bushes, fences, sheds and other flow obstacles.

The last simulation proved the anticipated fact, that the Kampa protection (part P0002) would hardly affect water levels near the Charles Bridge. The discharge through the Kampa Park is only 300 m^3/s out of 5,300 m^3/s of the total flood peak discharge. Therefore it was concluded that the influence on the flow conditions was negligible.

All the described simulation results were accepted by all the involved authorities and the 2D Prague Flood Model has become an integral part of the design process for the Prague Flood Protection measures. The evaluation of the final design of Prague Flood Protection has already started and no surprises are expected.

3.5. FLOOD PROTECTION – STRUCTURAL MEASURES

In the past, most cities and industrial installations in the Vltava and Elbe river basins were protected according to the formerly valid compulsory standards and norms for Q_{100}, Q_{50} or Q_{20} discharges. The structural measures were focused mainly on construction of reservoirs, improving river channel capacity and primarily designing dykes parallel to the river channel. Most of these measures cut-off flood plains and the assessment of their impact in the whole basin has not been done at all. Most of the hydraulic calculations were performed on the basis of steady non-uniform flow, which in many cases means an oversimplification.

In the current strategy, a new concept was introduced by the definition of the Flood Protection Strategy introduced by the Ministry of Agriculture. The most important highlights are:

- The flood management structures have to be assessed in the whole system and full detail, and only mathematical models would provide relevant and prompt assessment of impacts.
- Design floods are derived from historical floods including their dynamic characteristics. The impact of the newly proposed structures has to be compared with the current situation – with and without the structure installed.

- There has to be emphasis on enlarging the flood plain retention capacity along the river and any restriction of the flood retention volume should be compensated by other complementarily designed measures.
- Environmental aspects and hazards have to be assessed.
- Best forestry and agricultural practice have to be taken into account in order to increase the retention capacity in upper parts of the basin.
- Moveable (temporary) structures are very important structural measures; however, time and logistical arrangement required for their deployment at desired locations represent certain risks.
- A sediment transport has been ignored many times in the past and some structures have failed to operate properly due to this fact.
- Urban drainage systems (primarily sewers and local streams) directly impact on the functionality of flood mitigation structures and the assessment of mutual relationship between these two systems has to be performed in order to develop operational rules for the overall scheme.
- All newly proposed/designed structures have to be also assessed for the impacts of discharges considerably higher than the design flood. This is very important for flood risk management and contingency planning.
- All newly designed structures have to take into account emergency measures for events exceeding the design discharge (or water levels).

3.6. NON-STRUCTURAL MEASURES

According to the Water Act, the flood protection activity includes all the measures that protect human lives and property against floods. This aim should be achieved by preventive activities and increased retention capacity of the basin by systematic measures (e.g., new techniques in agriculture and forestry). This definition shows that the flood protection includes a wide range of activities, rather than just structural measures.

Non-structural measures include all activities, which help improve flow conditions in flood plains. This essentially calls for removal of all obstacles from the active parts of flood plains, including the clean-up of densely overgrown areas on the banks, islands, etc. In the 2002 Prague Flood Model, reducing the bed hydraulic resistance in critical areas has shown the highest effects among all the studied measures. Especially in the areas, constituting "bottlenecks" in some river reaches, vegetation on banks or islands was found to exert considerable effects (i.e., high resistance against water flow) preventing faster outflow from the reach and causing increased water levels and flooded areas upstream. In parts, where the whole valley is narrow and constrained, high bed resistance of the banks can cause water level increase in the range of some metres. In all the above mentioned cases, the 2D mathematical model is an extremely useful tool, an indicator, which can quickly locate the critical cross-sections in the river reach and show how much it would be possible to decrease the water level in the particular reach.

4. Conclusions

Flood events, which hit the Czech Republic in 1997 and 2002, were different and so were their consequences. In 2002, a considerably larger area of the Czech Republic was afflicted and the peak discharges were greater. Fortunately, it became clear that during the 2002 flood, some measures introduced after the 1997 experience helped save lives and property, namely in Prague. Authorities undertook several very important steps in the right direction; new laws were accepted and introduced into the daily life (Water Act, Law of Crises Management and Law of Integral Rescue System) and improved logistics and rescue operation organisation proved to be valuable during the 8/2002 flood. Communications were definitely better and co-operation among the Rescue Forces, Fire-brigades, Police and the Czech Army, was rated at a higher level than in 1997. Water management information service (such as flood mapping) was utilised for evacuation of about 220.000 inhabitants, and losses of life were very low during the flood. As expected, enormous losses of property were averted thanks to the newly constructed flood protection structural measures in Prague. Such measures proved themselves during the 8/2002 flood.

Finally, the 2002 flood fully proved the usefulness of mathematical models for flood protection at any stage, in design of flood protection measures, in improved flood forecasting, and in producing easily understood maps and charts enhancing the rescue activities.

Contingency plans and emergency scenarios for the adopted, designed and partly built structural measures will be prepared.

References

CHMI (2002) Preliminary report on hydrometeorological situation of the Flood in August 2002. The 2nd preliminary version of the report, August 29. CHMI, Prague, Czech Republic

DHI Hydroinform a.s. (2001) 2D Prague Flood Model. Study report, DHI Hydroinform a.s., Prague, Czech Republic

DHI Hydroinform a.s. (2003) 2D Prague Flood Model, version 2002. Study report, DHI Hydroinform a.s., Prague, Czech Republic

Hydroinform a.s. (1999) Prague Flood model: Comprehensive study. Study report, Hydroinform a.s., Prague, Czech Republic

Patera A (ed) (1999) How floods influence the development of Prague City. The Vltava River Basin Authority, Prague

SIMULATION OF THE SUPERIMPOSITION OF FLOODS IN THE UPPER TISZA REGION

JÓZSEF SZILÁGYI*, GÁBOR BÁLINT [1], ANDRÁS CSÍK, BALÁZS GAUZER, MARGIT HOROSZNÉ GULYÁS
„VITUKI" Environmental Protection and Water Management Research Institute, Budapest, Hungary

Abstract. Major floods on the Upper Tisza down to the confluence of the main river with the Bodrog tributary usually result from the superimposition of several flood waves arriving from upstream sections and their coincidence with floods on the tributaries. This phenomenon was the basis for simulation exercises which were carried out for a limited number of scenarios generated by the combination of a few historical events. A more complex approach using a hybrid seasonal Markov-chain model for daily streamflow generation was also applied in combination with the DLCM-based flood routing system of the complex river network. Diurnal increments of the rising limb of the main channel hydrograph, increments of the rising hydrograph values at the tributary sites, and the recession flow rates of the tributaries as well as of the main channel were identified and subject to various statistical analyses. The model-generated daily values retained the short-term characteristics of the original measured time series as well as the probability distributions and basic long-term statistics of the measured values. Such results describe possible future scenarios of flood events and may help water managers prepare for events that have not yet been recorded in the past, but could be expected in the future.

---

[1] To whom correspondence should be addressed. Gábor Bálint, „VITUKI" Environmental Protection and Water Management Research Institute, Kvassay Jenő út 1., Budapest, Hungary, H-1095; e-mail: balint@vituki.hu

* On leave from the Conservation and Survey Division, University of Nebraska-Lincoln, Lincoln, NE 68588-0517

Keywords: superimposition of floods, flood routing, streamflow generation models, Markov chains, hydrographs

1. Introduction

Hydrologists involved in operational stream forecasting and flood control may be interested in possible future scenarios of flood events. This may help them prepare for events that have not yet been observed in the past, but nonetheless could be expected in the future. While statistical analyses of e.g., annual maxima may yield information on return periods of floods of different magnitudes, they do not provide information on the possible time-sequence of the expected flood events. Such information may encompass duration of different water levels during flood, and the speed at which stream levels may rise or the flood may recede, all of which potentially influences the planning and organisation of flood protection works.

Major floods on the Upper Tisza down to the confluence of the main river with the Bodrog tributary (the Tokaj station) usually result from the superimposition of several flood waves from upstream sections and their coincidence with floods on the tributaries. This phenomenon was the basis for simulation exercises which were carried out for a limited number of scenarios generated by the combination of a few historical events (Bartha et al. 1998; Gauzer and Bartha 1999, 2001; Harkányi and Bálint 1997). These studies clearly indicated the possibility of unfavourable coincidences of floods; however, no estimates of frequency are associated with individual scenarios. Relatively short observed time series on the Tisza and tributaries (less than 150 years for water levels and less than 100 years for discharges) leave no space for attempts to include bi- or multi-dimensional distributions. To overcome these difficulties, a more complex approach using a Monte Carlo simulation for the upper boundary stations for daily streamflow generation, in combination with the DLCM based flood routing system used in the forecasting and modelling system of the National Hydrological Forecasting Service of Hungary, VITUKI, was used for the Upper Tisza river network (Figure 1).

Our multivariate, seasonal streamflow generation algorithm detailed below uses components of the shot noise models in a Markov-chain based approach together with a conceptual framework describing flow recession without the need for knowing precipitation. It is built around the concept of conditional heteroscedasticity, which was originally established in the ARCH models of time series analysis by assuming that the noise term was not independent of the process to be modelled or identically distributed (Engle 1982).

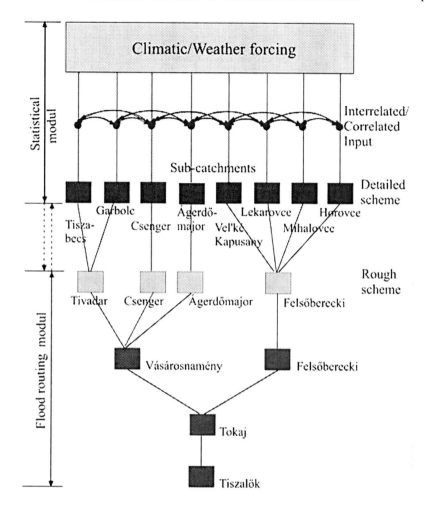

Fig. 1. General scheme for simulation of flood events on the Upper Tisza

2. Formulation of a hybrid Markov-chain model

The model works with daily streamflow data from which a time series of diurnal increments can be obtained by differentiating the original series. These increments define a two-state Markov chain for perennial streams. State one is observed when the increment is positive (termed as the "wet" state), and state two (termed as "dry") occurs when the increment is negative. The two states result in four different state transitions: wet-wet (P_{ww}), wet-dry (P_{wd}), dry-wet (P_{dw}), and dry-dry (P_{dd}). The state transition probabilities can be estimated from the observed data as

$$P_{ij} = \frac{n_{ij}}{\sum_j n_{ij}}; \quad i, j = w, d \qquad (1)$$

The state transition probabilities typically vary with seasons. Often these transitions are written on a monthly basis (e.g., Xu et al 2001, 2003), which results in very similar values between the neighbouring months, thus raising the question whether they are statistically different or not. From the viewpoint of parameter parsimony, a seasonal resolution should suffice in most cases, and this approach was adopted here.

2.1. RISING LIMB OF THE HYDROGRAPH

Positive diurnal increments (or wet states) designate the rising limb of the hydrograph. Sargent (1979) and Aksoy (2003) recommended a two-parameter gamma distribution for these increments. For the largest tributary of the Danube in Hungary, the Tisza River, and its tributaries, the Bodrog, Szamos, and Kraszna Rivers (Figure 1), a Weibull distribution fits well the observed data collected from 1951 to 2000.

During Monte-Carlo simulation of these increments (dQ [L^3T^{-1}]) for wet states the computer uses the fitted distributions (on a seasonal basis) for random number generation. The values obtained for the main channel, which is the Tisza River, are subsequently perturbed with an additional noise term (W [L^3T^{-1}]), taken from a normal distribution of zero mean (m). The noise, however, is not identically distributed, because its standard deviation is conditioned on the Weibull-distributed random number (dQ_{gen}) to be perturbed

$$W(m, \sigma) = W(0, a \cdot dQ_{gen}^b) \qquad (2)$$

where a [$L^{3(1-b)}T^{(b-1)}$] is a scale-coefficient, and b [-] is an exponent. The noise values that are negative and have larger magnitudes than the corresponding dQ_{gen} values, are discarded and replaced by zero. This makes the noise values to follow a positively skewed distribution, of which mean is no longer zero. The scale coefficient, a, and the exponent, b, are model parameters to be optimised. This so-called conditional heteroscedasticity assures that the Monte-Carlo generated diurnal increments will have a similarly wide range of values as the observed ones.

Another alternative could be the use of a noise term that follows the Weibull rather than normal distribution. In that case the prescribed mean and standard deviation of the distribution (the latter providing a rough idea of the spread of the distribution) could be converted to the parameters of the distribution. However, the Weibull distribution is a monotonic function for a wide range of parameters, while the distribution of the diurnal increments generally is not. Consequently, it is more convenient to employ a normal distribution instead, and make it skewed by specifying the lower limit with each value of the increment to be perturbed.

Once the positive increment values have been generated for a wet spell, they are ranked in an increasing order to make sure that the larger increments are closer to the peak of the hydrograph. This recreates the general shape and ensures preservation of the correlation structure of the rising limb of the hydrograph (Aksoy 2003).

Typically, tributary flow values are correlated with the main channel values; therefore, one may want to avoid generation of positive increment values for the tributaries separated from the main channel. One way of linking tributary increments to the main channel state could be achieved by conditioning the state transitions of the tributaries to that of the main channel, since for a correlated flow series the probability of a wet-to-wet transition is higher for the tributary when the main channel is in a wet state too. Unfortunately, such conditioning of the state transition probabilities did not meet expectations in our study: the cross-correlation value between the (measured) main channel and simulated tributary flow rates remained much lower than observed. As an alternative, the procedure described in the following section was performed.

Diurnal increments of the rising hydrograph at the tributary sites and at the main channel were described by second-order polynomials. The polynomial-derived tributary increment values were again perturbed by an additional noise term in eq. (2), similar to the main channel case. Note that, as before, eq. (2) includes dQ_{gen} of the main channel and not of the tributary. Alternatively, one may chose to use the polynomial-derived tributary value instead of dQ_{gen}. In either case, the coefficients, a and b, must be optimised for each tributary site.

2.2. RECESSION CURVE

Drainage of stored water from the channel is generally a nonlinear process (Aksoy et al. 2001). Often a nonlinear reservoir approach is used

$$Q = kS^n \qquad (3)$$

where Q $[L^3T^{-1}]$ is observed streamflow, k $[L^{3(1-n)}T^{-1}]$ is a storage coefficient, S $[L^3]$ is stored water volume, and n $[-]$ is an exponent. During recession flow the value of the exponent may change (Kavvas and Delleur 1984) with time. As an alternative, rather than changing n through time, the value of k may be changed (Aksoy et al. 2001; Aksoy 2003) with n chosen equal to one, and in this case, eq. (3) can be rewritten in a differentiated form as

$$\frac{dQ}{dt} = -kQ \tag{4}$$

which has a solution (up to an arbitrary constant)

$$Q(t) = Q_0 e^{-kt} \tag{5}$$

that can be written for $t = 1$ day and with the $Q_0 = Q(t-1)$ choice as

$$Q(t) = e^{-k'} Q(t-1) = c_1 Q(t-1) \tag{6}$$

where $k' (= 1\ k)\ [-]$. Employing a finite difference approximation of eq. (4) with $t = 1$ day yields

$$Q(t) = (1 - k')Q(t-1) = c_2 Q(t-1) \tag{7}$$

which shows that by proper choice of c_2 in the finite difference scheme one can obtain the analytical solution of eq.(6). By letting the value of k' in eq. (7) change in time, one can simulate the outflow of a nonlinear reservoir having a time variable exponent. The following expression permits the value of c_2 to increase in a logarithmic fashion from a minimum value at the time of the peak of the hydrograph to close to 1, if k'_{min} is chosen sufficiently small

$$Q(t) = Q(t-1)\left[1 - k'_{min} - \frac{k'_{max} - k'_{min}}{\ln\left(\frac{Q_{max}}{Q_{min}}\right)} \ln\left(\frac{Q(t-1)}{Q_{min}}\right)\right] \tag{8}$$

Note that when $Q(t-1) = Q_{max}$, $c2 = 1 - k'_{max}$; and when $Q(t-1) = Q_{min}$, $c2 = 1 - k'_{min}$. Eq. (8) assures that the recession is steeper than a negative exponential function, and so it fits observations (Kavvas and Delleur 1984).

The above description of recession flow cannot account for year-to-year variations in the volume of groundwater stored in the catchment. During wet years this additional source of water will prevent very low flow rates in the channel for perennial streams. The model accounts for this variability by adding a stochastic groundwater component to the recession flow model descried by eq. (8) in the form

$$Q_{gw}(t) = (1 - k'_{min})Q_{gw}(t-1) \qquad (9)$$

where Q_{gw} designates the groundwater contribution to the channel flow, which is the sum of eqs. (8) and (9) during recession flow periods. The starting value of Q_{gw} with a wet-to-dry transition at time t is obtained as

$$Q_{gw}(0) = \left| W(d \cdot Q_{gen}(t), f \cdot Q_{gen}(t) \right| \qquad (10)$$

where, again, d [-] and f [-] are parameters to be optimised; and W is a normally distributed variable. The straight brackets denote the absolute value. In theory the multiplier of Q_{gw} in eq. (9) could change with time as in the channel flow case (Brutsaert and Nieber, 1977; Szilagyi 1999, 2004), but that would further complicate the model which is intended to be kept as simple as possible.

The model has altogether 8 parameters to be specified. Two of them, Q_{max} and Q_{min}, are the observed extremes and can be specified during Monte-Carlo simulation to be somewhat larger and smaller, respectively, than their historical values, in order to accommodate possibly larger or smaller generated values than observed. From the remaining six parameters only a is dependent on the season and even then only for the main channel. The remaining parameters are constant during the year.

3. Model results

The Tisza River is the major tributary of the Danube in Hungary (Figure 1). Besides the gauging station of Tivadar on the Tisza River, three additional sites on different tributaries of the Tisza were included in this preliminary stage of the study. Fifty years (1951-2000) of daily instantaneous flow-rate values were employed for all four gauging stations representing the upper limit of the 'coarse scheme' (Figure 1). Table 1 displays the estimated state transition probabilities at Tivadar on a seasonal basis. It shows that a wet-to-wet transition has the highest likelihood in spring, which comes from two sources: (a) it is the season of most abundant precipitation in Hungary; and (b) it is the time of year when melting snow in the Carpathian Mountains feeds the streams, occasionally (especially when combined with rain) causing major flooding in the region. The positive diurnal increment values at Tivadar were fitted with Weibull distributions for each season and randomly generated using those distributions. For each increment a W value (eq. [2]) was added, with optimised values of a (= 1.1, 1.2, 1, 0.7, for the four seasons, starting with winter) and b (= 1). For each wet spell these values were sorted in an ascending order. Figure 2 displays the Q-Q plots of observed and generated positive diurnal increment values for the four seasons.

Table 1. Estimated state transition probabilities (%) at Tivadar

	P_{dd}	$P_{dw}\ (=1-P_{dd})$	P_{wd}	$P_{ww}\ (=1-P_{wd})$
Winter	80.44	19.56	37.9	62.1
Spring	79.71	20.29	37.55	62.45
Summer	76.15	23.85	51.34	48.66
Fall	80.51	19.49	48.62	51.38

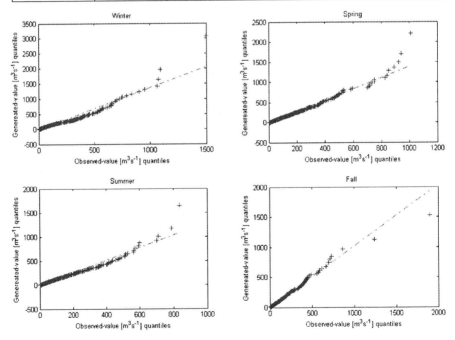

Fig. 2. Seasonal Q-Q plots of observed and generated positive diurnal increment values of the Tisza River at Tivadar

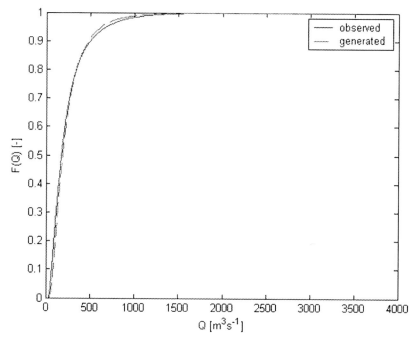

Fig. 3. Empirical cumulative distribution functions of 50 years of observed and simulated daily flow rates of the Tisza River at Tivadar

Recessions were modelled with $k'_{max} = 0.33$, $k'_{min} = 0.015$, $d = 0.04$, and $f = 0.02$. Comparison of observed and generated time series of daily discharges at Tivadar proves that the asymmetric shape of the observed hydrographs is well preserved in the generated data. Empirical cumulative distribution functions of 50 years of observed and simulated daily flow rates are demonstrated in Figure 3. Distributions of the annual maxima, means, and minima are also well preserved (Figure 4). A good agreement between observed and simulated flow rates for each season, the annual change in the median values (i.e. elevated water levels in spring, low flows in autumn), and the skewness of the rates was found. Finally, Figure 5 shows the autocorrelation functions for 50 years of observed and simulated daily flow values of the Tisza River at Tivadar.

Simulation of the tributary flow differs from that in the main channel only in the application of a polynomial regression between the tributary and main channel diurnal increments during wet spells of the main tributary in place of a Markov approach of state transition probabilities.

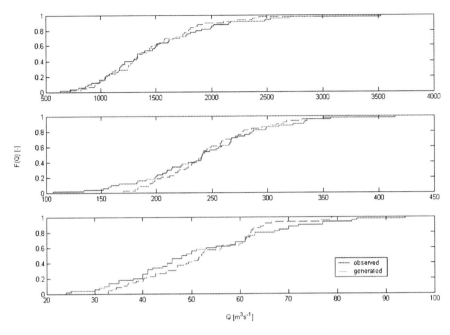

Fig. 4. Empirical cumulative distribution functions of 50 years of observed and simulated annual maxima, means, and minima of daily flow rates of the Tisza River at Tivadar

In summary it can be stated that by the application of the proposed hybrid, seasonal Markov-chain approach to daily flow simulation at multiple catchment sites it is possible to generate arbitrarily long time series of daily flow rates that fairly well preserve basic long-term (mean, variance, skewness, autocorrelation structure, and cross-correlations) statistics as well as the short-term behaviour (asymmetric hydrograph) of the original time series. The approach is centred on the concept of conditional heteroscedasticity which means that the noise term of the stochastic model applied is not independent of the process to be modelled, and it is not identically distributed. The model has altogether 9 parameters (in a seasonal formulation) for the main channel site to be optimised, and 6 additional parameters for each gauging station to be included. While the described approach is very simple, optimisation of the parameters may require some effort from the modeller.

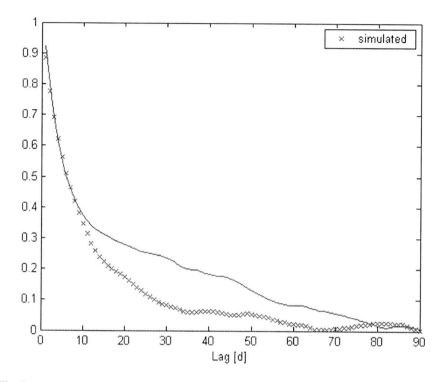

Fig. 5. Autocorrelation functions of 50 years of observed and simulated daily flow values of the Tisza River at Tivadar

Acknowledgment

This work has been supported by the Hungarian Research and Development Project: "Flood Risk Analysis", NKFP 3/067/2001.

Disclaimer

The views, conclusions and opinions expressed in this paper are solely those of the authors and not the University of Nebraska, State of Nebraska or any political subdivision thereof.

References

Aksoy H (2003) Markov chain-based modeling techniques for stochastic generation of daily intermittent streamflows. Adv. Water Resour. 26:663-671

Aksoy H, Bayazit M, Wittenberg H (2001) Probabilistic approach to modelling of recession curves. Hydrol. Sci. J. 46(2):269-285

Bartha P, Bálint G, Gauzer B (1998) Árvízi szimulációs vizsgálat a felső-tiszai árhullám forgatókönyvekre (Flood simulation study fort Upper Tisza flood scenarios) VITUKI report November

Brutsaert W, Nieber JL (1977) Regionalized drought flow hydrographs from a mature glaciated plateau. Water Resour. Res. 13(3):637-643

Engle RF (1982) Autoregressive conditional heteroscedasticity with estimates of the variance of UK inflation. Econometrica 50:987-1007

Gauzer B, Bartha P (2001) Árvízi szimulációs vizsgálatok a Tisza Tokaj-Szeged közötti szakaszán (Flood simulation experiments in the Tokaj-Szeged reach of the River Tisza) Vízügyi Közlemények, 2001(4):511-538

Gauzer B, Bartha P (1999) Az 1970. és 1998. évi felső-tiszai árhullámok összehasonlítása, árvízi szimulációs vizsgálatok. (Comparative simulation analysis of the 1970 and 1998 floods of the Upper-Tisza river) Vízügyi Közlemények 1999(3):354-390

Harkányi K, Bálint G (1997) Szélsőséges csapadékból kialakuló feltételezett árvízi helyzet vizsgálata (Analysis of an extreme rainfall generated flood scenario). Vízügyi Közlemények, 1997(4)

Kavvas ML, Delleur JW (1984) A statistical analysis of the daily streamflow hydrograph. J. Hydrol. 71:253-275

Sargent DM (1979) A simplified model for the generation of daily streamflows. Hydrol. Sci. Bull. 24(4):509-527

Szilagyi J (1999) On the use of semi-logarithmic plots for baseflow separation. Ground Water 37(5):660-662

Szilagyi J (2004) On heuristic continuous baseflow separation. J. Hydrol. Engin. 9(4):1-8

HARMONISING QUALITY ASSURANCE IN MODEL-BASED STUDIES OF CATCHMENT AND RIVER BASIN MANAGEMENT

JAN SPATKA[1]
DHI Hydroinform, Prague, Czech Republic, 100 00

Abstract. This paper presents the latest results of a nearly finished WP5 project named "HarmoniQuA". This international project deals with harmonisation of quality assurance in model-based studies of river basin management, and consists of four work packages, focused on knowledge base development, supporting tool development, tool testing and dissemination. As a general result of this project the MoST (Modelling Support Tool) is being produced, which combines the knowledge base and a guideline tool.

Keywords: Quality assurance, mathematical modelling, water management, water framework directive, good modelling practice

1. Introduction

The methods of mathematical modelling are currently very popular and daily used in all branches of the water management sector. It is well known that relatively small investments into a detailed study of impacts of proposed actions or measures on the hydrological system are always positively reflected by savings on investment costs for the implementation of such measures.

---

[1] To whom correspondence should be addressed. Jan Spatka, DHI Hydroinform, Na vršich 5, 100 00 Prague 10, Czech Republic; e-mail: j.spatka@dhi.cz

Nowadays, it is possible to apply mathematical models to different parts of hydrological cycle either separately or globally, depending on the selected mathematical model:

- Precipitation-runoff models (forecast services, impact of land use changes, generation of design discharges)
- Hydrological models of surface waters (river and flood plain flow studies, flood models, operation of reservoirs, water quality studies, sediment transport studies, pollution transport studies, identification of flood plains and active zones, minimum discharge definition, forecasting models, models of costal zones)
- Hydrological models of groundwater (models of unsaturated and saturated groundwater zones, groundwater quality models, groundwater quantity models, groundwater pollution transport models
- Detailed 3D hydraulic models (detailed studies of hydraulics of structures, studies of flow around obstacles, detailed groundwater studies of the saturated zone, detailed studies of costal zones and harbours)
- Global hydrological models (global studies of land use changes, conceptual models of hydrological cycle changes)
- Socio-economic models (risk analysis, cost-benefit analysis), and
- Biota models (studies of biota development, landscape studies, ecological studies).

In any case, the study costs are relatively small, in relation to the investment costs needed for practical implementation, but not negligible. Unfortunately and contrary to the design implementation phase, where several international and uniform quality control standards already exist, there are no such instruments widely applicable to mathematical modelling studies. Although some local or even national standards or recommendations exist, they are mostly non-obligatory, cover only one or a limited number of water sector domains at a local or national level and as such, they can be used only as non-binding guidelines in other countries (if they are even known there). Furthermore, these guidelines differ throughout Europe. The consequences are that the resulting modelling studies and the decisions based on them often are non-transparent, irreproducible, non-auditable and not fully comparable among different countries.

The Water Framework Directive (WFD) provides a European policy at the river basin scale. In the WFD it is explicitly mentioned that water resources models should be used to support the planning process and the decision making.

Thus, the needs for improvement can be summarised as follows: standardisation and harmonisation to allow the auditing of modelling projects; improved quality of modelling and simulation by using more homogenous and better software; methodologies and software enabling quantification of uncertainties and sensitivities; integration of hydrological, socio-economic and legal-political issues; and, tools for evaluation of the appropriate use of models. To fulfil these needs, the HarmoniQuA project was initiated as a part of the 5th Framework Program. The main goal of the HarmoniQuA project is to fill the aforementioned gaps and supply a missing chain link in the Water Framework Directive.

2. Description of HarmoniQuA

Harmonising **Qu**ality **A**ssurance in model-based catchment and river basin management projects (**HarmoniQuA**) is an international project conducted under the 5th Framework Program dealing with quality assurance standards of mathematical modelling in the water management sector. This project is conducted by a large consortium of international experts from universities, research institutes and consulting companies all over Europe, including: the Wageningen University (The Netherlands), Geological Survey of Denmark and Greenland (Denmark), National Technical University of Athens (Greece), Centre for Ecology and Hydrology (United Kingdom), WL/Delft Hydraulics (The Netherlands), Cemagref (France), Bundesanstalt für Gewässerkunde (Germany), Swedish Meteorological and Hydrological Institute (Sweden), VITUKI Plc (Hungary), University of Dortmund (Germany), Laboratório Nacional de Engenharia Civil (Portugal), and DHI Hydroinform a.s. (Czech Republic).

The overall goal of HarmoniQuA is to improve the quality of model-based river basin management and enhance the confidence of all stakeholders in the use of models. This main goal will be reached by providing a methodological layer in the modelling infrastructure. The HarmoniQuA project upgrades methodological expertise and identifies and fills gaps in it. The knowledge on modelling and simulation consists of the generic knowledge and domain-specific and software-specific aspects, in combination with a clear glossary of terms and concepts.

This body of knowledge is archived in a knowledge base, in order to facilitate its use and upgrading. A software tool is developed and produces guidelines to help modellers, managers and other stakeholders throughout a model-based water management study. These guidelines are specific for the type of application (planning, design, or operational management) and specific for the complexity of the task (routine, professional, or research). The following parts of the tool monitor all the tasks performed in such studies and store model project summaries in a database. These summaries use a similar data structure as the knowledge base. Another part of the tool enables domain, user, application and job type specific reporting of all tasks and activities performed in the study. Finally, the tool can learn from these monitored modelling histories and use such experiences in future projects.

Dissemination and exploitation infrastructure would provide long term support and guarantee future use by the entire community of modellers. This is achieved by setting up an infrastructure for dissemination and exploitation for project benefits, designing, preparing and providing courseware to train new professionals in good modelling practices; organising workshops and discussion sessions with professional bodies and practitioners to facilitate moving towards harmonisation and standardisation for modelling in water management at the European level; and, opening up the concept of quality-controlled modelling to any interested stakeholders including planners, policy-makers and concerned members of the public.

2.1. HARMONIQUA OBJECTIVES AND DELIVERABLES

HarmoniQuA has four specific objectives, which are solved in four work-packages: (a) to develop a generic, scientifically based methodology and a set of guidelines for the modelling process; (b) to develop tools to support modellers and water managers throughout the quality assurance process; (c) to test the methodology/guidelines and the support tools in actual cases; and, (d) to disseminate the results of the project to users in the academic education sector, water managers and model users, and other interested stakeholders.

The project has the following deliverables: (a) one shared European methodology on model-based water management covering generic and specific aspects for single domains, multi-domains and levels of integration at a river basin scale, structured in a knowledge base (3 versions); (b) a set of tools enabling to use this methodology and support the entire process of modelling and simulation (3 versions); (c) results of two sets of real-life case studies; and, (d) an infrastructure for exploitation and dissemination of project results, including courseware.

3. The HarmoniQuA Modelling Support Tool (MoST)

The Harmoniqua project divided the water sector into the following domains, which could either singularly or in combination (multi-domain projects) cover all potential types of water studies: Biota, Flood Forecasting, Groundwater, Hydrodynamics, Precipitation-Runoff, Socio-Economic, Water Quality, and Generic – common tasks for all the above

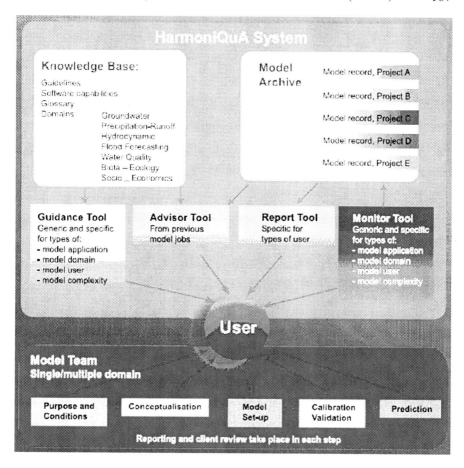

Fig. 1. Schematics of the HarmoniQuA system

domains. This is also the structure used for the HarmoniQuA Knowledge Base and the HarmoniQuA tool organisation.

The HarmoniQuA project has developed the Modelling Support Tool (MoST) to provide a user-friendly guidance and quality assurance framework that will contribute to enhancing the credibility of catchment and river basin modelling. It prompts users with the appropriate 'next step' commands in the modelling process and provides an audit trail to check previous decisions. MoST has further evolved since it was initially tested in the winter of 2003. A Knowledge Base containing knowledge specific to seven domains - groundwater, precipitation-runoff, river hydrodynamics, flood forecasting, water quality, ecology, and socio-economics - forms the heart of the tool.

MoST posses the following functionalities serving to:

- Guide: to ensure that the model has been properly applied;
- Monitor: to record decisions, methods and data used in modelling tasks; and,
- Report: to provide reports suitable for managers/clients, modellers, auditors, stakeholders and the general public.

MoST now supports multi-domain studies, accommodates several types of users (water managers, modellers, auditors, stakeholders and members of the public), and contains an interactive glossary that is accessible via hyperlinked text. The second round of testing (autumn 2004) focused on testing the application of MoST in complex multi-domain modelling studies.

3.1. BREAKDOWN OF THE MODELLING PROCESS

The modelling process has been divided into five steps, as shown in Fig. 2. Each step includes several tasks, which in turn involve a range of activities. The later steps end with a reporting task and a client review of past progress and future plans.

A computer-based journal is produced within MoST where the water manager and the modelling team record the progress and decisions made during the study according to the tasks in the flowchart (Fig. 2).

3.2. USING HARMONIQUA MOST

After specifying 'New Project', the next stage in using MoST is to define subprojects (see Fig. 3). A subproject is a clearly defined domain-specific or integrated multi-domain component of a modelling study. The user then selects the necessary tasks (Fig. 2), adds users and their authorisations (Fig. 4), and edits the scoreboards, which are used to audit the quality of modelling work. The screen then changes giving access to the guidelines, project and reporting components of MoST. The guideline and project components are built in three re-sizeable and scrollable panels showing tabbed or menu linked pages (see the example of Project Components in Fig. 5). A hyperlinked glossary is active in both screens. For additional and better readable screen pictures, please visit the project website: www.HarmoniQuA.org.

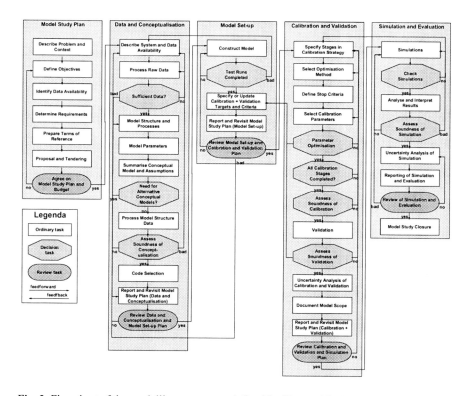

Fig. 2. Flowchart of the modelling process as defined by HarmoniQuA

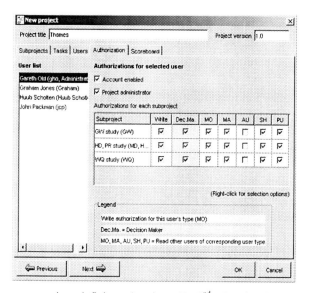

Fig. 3. Starting a new project: defining subprojects (the 2nd subproject is multi-domain)

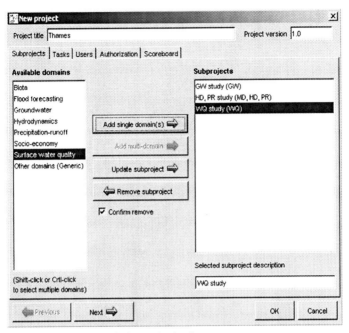

Fig. 4. Starting a new project: Defining authorisations

3.2.1. *Guideline component*

The left-hand panel displays a tree view of model steps and tasks. This information is displayed in a flowchart in the upper right panel. In the lower right panel task descriptions and information on activities and methods are presented. The user may browse guidance for future activities without changing their current position in the project component.

3.2.2. *Project component*

User activities are monitored and written to a model journal, which is saved on the local computer or a server, allowing other modelling team members to record their work in the same model journal or to read what others have done. A tree view of tasks is displayed in the left-hand panel. The upper-right panel shows the journal information for a task selected from the left-hand panel (e.g. Model Parameters in Fig. 5). Initially, the title page of the task is shown, with tick boxes to identify the task status, and a scrollable menu showing the list of relevant activities and their status. In fully synchronised multi-domain studies, in which two or more single domains are treated as an integrated single domain (e.g. second subproject in Fig. 3), domain specific versions of the same activity often occur. In other studies domain specific subprojects are undertaken in parallel by different teams at different speeds (e.g., see the list of subprojects in Fig. 3).

Selecting an activity from the scrollable list brings up the relevant journal fields together with any existing entries (e.g. Parameters not calibrated, see Fig. 5). New information can be added on modelling activity, the outcome, methods applied and the reports or data sources used.

The user may then 'tick' the activity (and task) as completed, or could skip to another activity. Guidance for the selected task is presented in the bottom right panel. In the sample screenshot below (Fig. 5), 'Activity information' has been selected and scrolled to show the relevant guidance.

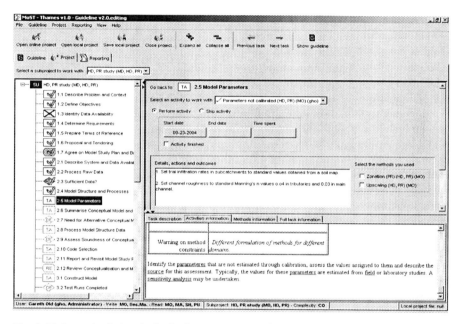

Fig. 5. Main work window in the Project component of MoST: Monitoring of tasks

3.2.3. *Reporting component*

This component enables users to tailor reports by specifying tasks, defining task and activity fields, and selecting user type filters. Although the modelling process consists of a fixed task structure, the system is flexible. For example, a project administrator may specify the tasks that must be completed by the modeller.

4. Dissemination

By the end of 2004 it was possible to download/view the following from the project public website (www.HarmoniQuA.org):

- The most recent full version of MoST;

- A series of computer screen movies and presentations that have been developed to introduce professionals and students to HarmoniQuA MoST; and,
- A programme of training sessions on how to use MoST to be held in 2005.

It is expected that the HarmoniQuA project and its results can contribute significantly to improving the quality of modelling and harmonising modelling practises throughout Europe. Such harmonisation of river basin modelling procedures in relation to the WFD will stimulate the compatibility and the consistency in modelling and will thus effectively support the development of a consistent water management policy across Europe. Finally, this harmonisation will open up the concept of quality-controlled modelling to any interested stakeholder including planners, policy-makers and concerned members of the public.

Acknowledgement

This paper is presented on behalf of the HarmoniQuA project consortium (named in Section 2) and used the actual project results.

References

HarmoniQuA project public website: www.HarmoniQuA.org

CHAPTER 3   FLOOD MANAGEMENT

RESEARCH, EDUCATION AND INFORMATION SYSTEMS IN THE CONTEXT OF A FRAMEWORK FOR FLOOD MANAGEMENT

ROLAND K. PRICE [1]
Professor of Hydroinformatics, UNESCO-IHE, Delft, The Netherlands

Abstract. The proposed framework of the International Flood Initiative for flood management is based on monitoring and observation of flood generation, occurrence and post recovery, data archiving and analysis, modelling, and systems for the support of decision-making. Guiding principles for the Initiative include living with floods, inter-disciplinarity bringing different skills together at the river basin level, empowered participation of all stakeholders, international and regional cooperation that is especially needed for trans-boundary catchments and rivers, and trans-sectorality that brings together the diverse knowledge generated in other sectors concerning floods. New research results are accumulated on a theoretical analysis of flood propagation including treatment of lateral inflows and modelling error, on the reliability of flood prediction and forecasting for improved risk assessment and reduction, on integrated flood management involving modelling support, and on the use of distributed decision support systems for transparent and empowering participation of stakeholders and the public. These results contribute to a range of different initiatives.

Keywords: decision support, floods, information, management, modeling, risk, uncertainty

---

[1] To whom correspondence should be addressed. Roland K Price, UNESCO-IHE, P.O. Box 3015, 2601 DA Delft, The Netherlands

1. Introduction

The situation regarding floods and the damage they cause is serious for the world economy and for the millions of people who are affected annually. Those who suffer most from the devastating effects of floods live in the least developed countries: casualties are high, economic losses are a significant proportion of the countries' GDPs, and the social disruption can be overwhelming. The world community has long been aware of the need to generate and transfer knowledge about systems and practices to reduce risks associated with floods. These however have focused on the introduction of engineering technology at the expense of the social, cultural and other reasons why people live in areas at risk of flooding. There is an urgent need for better, more effective and integrated methods to prepare for and mitigate flood disasters. The shortage of funding and political good will exacerbates the need for improved coordination at all levels.

The need to take international initiatives to alleviate the serious situation regarding flood disasters is high on the agenda of UN agencies. At the 16th session of the Intergovernmental Council of the International Hydrological Program (IHP) (September 2004) it was recommended (Resolution XV-14) that an International Flood Initiative/Programme (IFI/P) be launched as an interdisciplinary activity to mitigate the impacts of flooding worldwide, in cooperation with the World Meteorological Organisation (WMO) and other relevant UN agencies and NGOs. The intention is that IFI/P will develop into an interagency programme led by UNESCO. This initiative has synergy with a number of other goals and objectives of the UN and its agencies. It will contribute to meeting the relevant Millennium Development Goals. It will have an input to the UN Decade on Water for All (2005-2015). It will form part of the follow-up to the World Conference on Disaster Reduction held in Kobe, January 2005. It will be a means of implementing the UN Decade on Education for Sustainable Development (2005-2014).

The IFI/P will be a vehicle to carry forward and implement the commitment of the international community to vulnerability and risk management of floods, including prevention, mitigation, preparedness, response and recovery. The intention is that IFI/P will grow into a global programme with multiple national and regional activities.

At the heart of IFI/P is a focus on integrated flood risk management. Here risk is a function of the flood hazard and its potential impacts. The aim is to reduce the human and socio-economic losses caused by flooding while at the same time taking into account the social, economic, and ecological benefits from floods and the use of flood plains. Care is taken to integrate land and water resources development, including the institutional components of flood risk management, and to recognise the critical importance of stakeholder participation. In general IFI/P aims at implementing recommendations arising from the World Summit on Sustainable Development (Johannesburg 2002) and the World Conference on Disaster Reduction (Kobe 2005), considering the physical aspects of flooding, its socio-economic conditions and the risk a society is prepared to take in order to achieve its development objectives. The IFI/P involves a paradigm shift from the traditional, fragmented approach to flood risk management to a more holistic methodology.

2. Guiding principles

IFI considers six guiding principles:

- *Living with floods.* This principle recognises floods as inevitable natural phenomena. Whereas there are some beneficial aspects of floods, their negative impacts can be reduced through an understanding that risks result from a combination of flood hazards and societal vulnerabilities. Hazards can be reduced through appropriate physical interventions and the introduction of various technologies for warning and control. Similarly, improved planning, social awareness and preparedness, legislation, and a number of other factors can affect societal vulnerabilities. Above all there needs to be a concerted effort to reduce risks through a holistic approach to flood management. Communities and governments should be assisted, through proactive multi-hazard and risk-based approaches, to develop culturally sensitive and sustainable flood risk management strategies that harmonise structural and non-structural measures.

- *Equity:* The burdens and benefits of flood risk management (prevention, response and recovery) must be equitably distributed. Such a distribution has both ethical and legal dimensions. It is inevitable that flood risk management promotes appropriate policy processes and outcomes that are viewed as fair and legitimate among all the affected parties or stakeholders. This is also a sustainability issue such that the stakeholders should include future generations. Flood risk management strategies must therefore promote intergenerational equity.

- *Empowered participation.* Participation of all stakeholders, including individuals and communities, is widely recognised as a crucial factor for the successful implementation of flood risk management. Such participation becomes empowered only through appropriate institutional frameworks and innovative governance mechanisms, as well as through carefully designed enabling systems and environments using modern communication technologies that coordinate flood related activities at all levels. Flood risk management has to be part and parcel of social development.

- *Inter-disciplinarity*. Flood management is a process involving technical, social, ecological, economical, legal and political skills. There is an urgent need to develop and enhance the integration and communication between knowledge systems necessary to facilitate all flood related activities. These include physical aspects such as identifying and monitoring data from the real world to generate the necessary information for decision making in all areas, improving the statistical and other inductive forms of analysing flood data, and developing robust and appropriate flood modelling and real time forecasting and warning systems. The integration of appropriate skills is best achieved at the river basin level. All relevant natural, economic and social scientific knowledge together with appropriate technologies for data and information acquisition and communication should be brought together. The involvement of disciplinary stakeholders is also essential to ensure that the right institutional structures, participatory processes and policies are in place to promote fair and effective flood risk management.

- *Trans-sectorality*. All stakeholders need to be involved in flood risk management. Links should be established between the scientific community, decision-makers inside and between governmental levels, the relevant UN bodies, national and international organizations, NGOs, and market actors. This multi-sectoral approach would increase the effectiveness of processes and the acceptance of flood risk management decisions, and therefore enhance their sustainability.

- *International and regional cooperation*. This principle recognises the imperative for a cooperative, partnership approach for flood-related issues, especially when dealing with transboundary catchments and rivers. International and regional cooperation is also needed for transferring appropriate technologies for flood management, sharing of experiences in flood management, and for making the most beneficial local use of global systems for monitoring, observation and prediction of flood-generating natural events. Data, information and knowledge exchange and management are facilitated through cooperative networks, such as the IHP National Committees, UNESCO Water Centres, IAHS National Committees and other UN institutions and initiatives. The development, promotion and transfer of appropriate technologies in flood risk management are also a matter of international cooperation. Particular attention needs to be given to the specific problems of the poorest developing countries to enable them to cope effectively and fairly with flood hazards as a part of their national strategies for poverty alleviation.

3. IFI framework for flood management

In order to achieve its objectives IFI will promote an integrated and holistic framework for dealing with flood management problems. Such a framework will permit the integration of natural, scientific, economic, social, institutional and legal aspects of flood management. As such the framework would make a substantial contribution to facilitating the exchange of data and information on historical and real time flood management at different operational levels in transboundary rivers. This hydroinformatics framework can be summarised as having the following components:

- *Monitoring and observation* of flood generation, occurrence, and post recovery. The data collected from the natural system need to be combined with data on physical, economical, and social aspects of both historic floods and preparedness for future floods including descriptions of available flood warning and management plans. These details are needed to maximise the benefits of all available information including global observations from satellite imagery, regional and local hydrometeorological data on historical and real time flood events, and flood management experiences at different levels.

- *Archiving and analysis* of all collected data. Archiving needs to be done in globally accessible databases and information generating systems to provide useful data for early warning on flood generating events to all developing countries, and particularly to the most vulnerable sectors of their societies, via the installation of adequate information translation and dissemination systems.

- *Modelling* can be classified into three categories, depending on the purpose of the modelling activities. The first category concerns the modelling of the natural system for improved flood forecasting and prediction, leading to the development of flood warning systems and operational strategies during floods. The second category includes integrated models for the development of flood management plans (covering the planning, operational and post recovery phases of flood management), standards and legislation, in which the models of the natural system are combined with models of economic assessment, vulnerability and risk assessment, as well as environmental and social impact assessment. The third category focuses on the modelling of long term natural changes (including climate), coupled with parallel population and social changes for the purposes of corresponding scenario analyses, which will lead to long term strategies for coping with floods.

- *Decision Support Systems (DSS)* are needed to empower the participation of all stakeholders and the general public. These DSSs will enable participants in the decision-making or assessment process to access relevant knowledge, allowing them to express their interests and concerns through the formulation of judgements about the appropriate courses of action, and to be aware about the 'social landscape' or the interests and positions of other participants in the process. In addition, these systems should form the basis for meaningful collaboration and negotiation of all participating parties. The DSSs envisaged are systems that are of a socio-technical nature and involve both adequate institutional arrangements and an appropriate ICT implementation. The models described in the previous component (particularly those of the second category) will be at the core component of these DSSs.

4. Proposed IFI activities

4.1. RESEARCH

Out of the many research interests three aspects should be emphasised:
- Improved reliability of flood forecasting and prediction, and its contribution to improved risk assessment and reduction.
- Integrated flood management with modelling support, for improved risk allocation and management, flood management planning and standardisation.
- Design and development of distributed DSSs for transparent and empowered stakeholder and public participation in flood management.

4.2. EDUCATION

Educational activities will also be organised in the research areas described above. In addition to the post-graduate education at Masters level, a number of distance-learning short courses will be developed, on the following topics related to flood management:
- Flood modelling for management
- Integrated river basin management
- Structural and non-structural measures for flood management
- Flood management planning

4.3. INFORMATION NETWORKS AND SYSTEMS

Information networks form an important component of the programme:

- Development of a global database on extreme flood events, together with the description of physical, economic and social aspects of historic floods
- Open access and a reviewed repository of proven methodologies and tools for flood forecasting, prediction, analysis and management
- Open access to a multi-lingual CDS/ISIS based international bibliographical database on flood literature and reports
- Network between IFI and relief and humanitarian agencies.

4.4. EXPECTED RESULTS

The expected results are as follows:

- Improved understanding of complex flood defence systems, their failure modes and their interaction with natural processes
- Improved understanding of vulnerability of the public and assets to flood damage
- Improved disaster preparedness, evacuation and emergency management procedures and social resilience
- Incorporation of the integrated holistic framework into management practice
- Consequent reduction in human flood casualties and economic loss

There are a large number of associated key issues that can be highlighted in the context of the IFI/P. The remainder of this paper focuses on a few issues that have wide ranging repercussions. The first is to do with the information required for safe and reliable decision-making in flood risk management, and the second deals with the dissemination of established information through education.

5. Information for flood risk management

A river flood is the inundation of normally dry land (or wetlands) adjacent to a channel due to inadequate capacity of the channel to convey the current flow. Any analysis of flooding for planning or real time warning purposes is dependent on the propagation of information from upstream to downstream. The issue of the estimation or prediction of flood discharges and levels is a function of how excess rainfall-runoff accumulates down the dendritic stream network and of the interaction of the accumulated flow with the local geometry. What makes the calculation of flooding so difficult and intriguing are the wide variations that occur at the different stages: in the timing and distribution of rainfall, the complex interacting processes that determine what water runs off over the surface and what infiltrates into the ground, the non-linear behaviour of the surface runoff and the flow in the ground through the saturated and unsaturated zones, the collection in the streams, and the non-linear propagation of the accumulating flow along the stream network.

Flood prediction whether for planning or forecasting in real time is fraught with difficulty if only because of the complexity of the catchment and the hydrological and hydraulic processes involved. Add to these human interference in bringing about long term changes to the catchment in terms of land use and so-called flood protection schemes, or in generating short term operational events at dams, or even to the extent of deliberately breaking dykes to inundate adjacent land, then the possibilities for uncertainty in knowing what discharges and levels will be achieved during a flood event are considerable. The stochastic nature of flood events and their uncertain behaviour, at least in the details, raises the notion of risk as an additional measure in the context of a flood event, and of minimising the risk by having an adequate supply of relevant data from which the necessary information can be extracted. Inevitably we are concerned with making decisions based on information derived from limited data. Given the need to make a decision we are concerned to secure the necessary information on flood levels and discharges (and timing in the case of forecasting), coupled with a minimum uncertainty or risk associated with the values of the dependent variables. Generally this implies making optimal use of all available data, or, if we are in a sufficiently fortunate position, putting in place a data-monitoring network that provides a sufficient data set from which the required information for the decision making can be extracted.

This raises the notion of the availability of data and the information that can be extracted. This has direct relevance to river basins, which are not contained within one administrative boundary, and where monitored data are only available to single organisations within the basin. We provide some examples to illustrate this point. Consider first that a river basin is divided between two or more administrative authorities. Suppose that the most downstream authority does not receive any data on what is happening in the basin from authorities upstream. The primary information that the authority has is from a gauging station at the entry point of the river into the authorities region and at other monitoring points downstream. This is a very data-poor situation and therefore information poor. So far as flood warning is concerned, the only information available is a time series at the upstream gauging station. Obviously there is a certain lag time between the gauging station and a site downstream. There is therefore a possibility to provide some limited warning. Any improvement in the warning has to come from an analysis of the time series at the gauge, or from remote sensing by satellite of the river upstream. Chaos theory methods have been used successfully to analyse discharge time series at a gauging station on the India-Bangladesh border to improve the lead-time in the forecasting of floods in Bangladesh; see Rahman (1999) and also Babovic and Keijzer (2000).

A second example concerns the availability of discharge time series at two gauging stations some distance apart. With two time series rather than one, we are in a much better position to put together a forecasting model at the downstream station than we were before. The problem is to determine the maximum amount of information from the two series for flood prediction and forecasting in particular. Put in this way the information content between the two time series can be explored using Average Mutual Information (AMI) or correlation. AMI has the advantage that it can deal with non-linear connections. It provides a measure of the strength of the mutual information and the time at which the maximum mutual information is achieved. Any appropriate structure can then be used to link together the two time series. This is a problem of training a suitable data driven model, such as an artificial neural network, a fuzzy logic generator, a model tree, a genetic programming tool, a nearest neighbour technique or a support vector machine; see for example, Solomatine and Dulal (2003), Solomon and Price (2004), Abebe and Price (2004). These structures have nothing to do with the physics of the problem (except in so far as the input variables are appropriately selected). As such they are not optimal in the sense that they do not make use of other, physically based knowledge available. For example, it can be shown that the propagation of a flood between two sites in a river can be described by an approximation of the 1D Saint Venant equations namely:

$$\frac{\partial}{\partial t}\int_0^Q \frac{dQ'}{c_0(Q')} + \frac{\partial Q}{\partial x} + \frac{\partial}{\partial t}\left(\frac{a_0(Q)}{c_0^2(Q)}\frac{\partial Q}{\partial x}\right) = q - \frac{\partial}{\partial x}\left(\frac{Qq}{2Bs_0 c_0}\right)$$

where $c_0(Q)$ and $a_0(Q)$ are non-linear functions defining the kinematic wave speed and attenuation parameter in terms of the discharge, and q is the uniformly distributed lateral inflow (B is the surface width, which is a function of discharge, and s_0 is the bottom slope along the reach), c_0 and a_0 are stationary functions that describe in an average sense the geometry (and roughness) of the river reach, including any inundated flood plains. Given c_0 and a_0, the upstream discharge, the initial discharge distribution along the reach and q it is a simple matter to predict the downstream discharge. However, c_0, a_0 and q are not known initially. The equation above can be solved implicitly for q given the upstream and downstream discharges (and the initial discharge along the reach) and preliminary forms for c_0 and a_0. But c_0 and a_0 are only estimated to begin with: they still have to be determined. Minimising the negative values of the calculated q can do this. The separation of the lateral inflow from the model error can be done using an averaging filter that has a period approximately equal to the time of travel of the flood along the reach. This produces a quasi-lateral inflow, which strictly should have no negative values. The optimisation minimises the sum of any negative values. If there are no negative values then the sum of the negative values of the quasi model error is minimised while ensuring that there are still no negative values of the corresponding quasi-lateral inflow. Alternatively the objective function can be the sum of the root mean square errors that accumulate from the mismatch between the predicted and observed discharges downstream when using the non-negative filtered lateral inflow values. To facilitate the optimisation particular parametric forms are deduced for c_0 and a_0 including a total of 9 parameters. These parameters are determined as part of the optimisation process. The ACCOL method in the global optimisation tool GLOBE was used; see Solomatine (1998). There remain questions about the procedure being sub-optimal, but the method works well.

See Figures 1 and 2 for the optimised curves for c_0 and a_0 for a reach of the River Wye, UK, compared with forms deduced directly from data; see NERC (1975). Figure 3 gives an example prediction of the discharge downstream. For further background information see Price (1984).

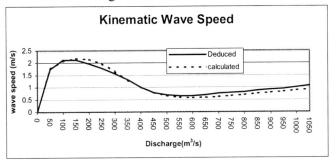

Fig. 1. Kinematic wave speed for the River Wye, UK

There are two further issues however. One is to determine the quasi-lateral inflow, and the other is to manage the quasi model error. The quasi-lateral inflow can be modelled with the local rainfall as input using an artificial neural network model, for example. The same can be done with the quasi-model error, this time using the upstream and downstream time series as input. Once the error is itself modelled it is possible then to determine bounds on any residual error to assess the uncertainty in the full model predictions. Alternatively it may be sufficient to deduce (variable) error bounds on the quasi-model error. Note that the provision of a measure of the uncertainty is important for the calculation of risk. For further information see Maskey et al. (2004) and Maskey and Price (2004).

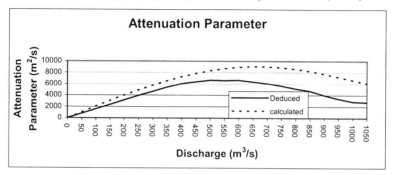

Fig. 2. Attenuation parameter for the River Wye, UK

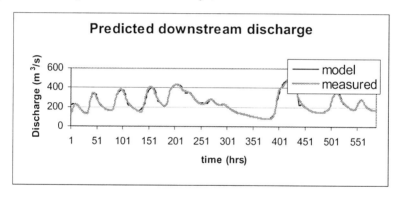

Fig. 3. Predicted discharge downstream for the River Wye, UK

A similar approach could be adopted for one or more rainfall time series and the corresponding outflow discharge time series from the catchment. The emphasis here would be on the calculation of the hydrological losses with, say, a lumped conceptual model.

The information space can be regarded as having three important dimensions. The first dimension includes data from the real world obtained through monitoring, remote sensing, personal observation and so on. The second dimension is all about the analysis of that data by a variety of means including statistical analysis, inductive modelling (data driven modelling), deductive modelling (such as computational hydraulics), systems analysis, and so on, in which human interpretation forms an important component. With these techniques we derive the necessary information for decision-making, along with assessment of uncertainty and risk, confidence bounds, and so on. Lastly, the third dimension includes the actual processes in which decisions are made by interacting groups of human agents. This draws on such tools as social actor theory, cybernetics, and decision theory, as well as being heavily dependent on a range of other methods in the social sciences.

6. Conclusions

This paper has presented briefly proposals for a new international initiative in flood research being put forward by UNESCO and WMO. It is intended to conduct the research within a hydroinformatics framework that addresses the acquisition and flow of information necessary for safe and reliable decision-making in the management of floods, especially in transboundary rivers. A huge range of information management tools is available. The objective should always be to make best use of all available information generated by efficient data collection, analysis, modelling, and decision support facilities, and including uncertainty and risk as vital components.

References

Abebe AJ, Price RK (2004) Information theory and neural networks for managing uncertainty in flood routing. J. of Computing in Civil Engineering ASCE 18(4):373-380

GLOBE http://www.data-machine.com

Maskey S, Price RK (2004) Assessment of uncertainty in flood forecasting using probabilistic and fuzzy approaches. In: Liong, Phoon, Babovic (eds) Hydroinformatics 2004: Proc. 6th Int. Conference on Hydroinformatics, World Scientific Publishing, Singapore, pp 1753-1760

Maskey S, Guinot V, Price RK (2004) Treatment of precipitation uncertainty in rainfall-runoff modelling: a fuzzy set approach. Advances in Water Resources 27(9):889-898

Natural Environment Research Council (1975) Flood routing studies, Vol. 3 of the Flood Studies Report, NERC, London

Price RK (1984) Flood routing in rivers, Chapter 4 In: Novak (ed) Developments in Hydraulic Engineering, Applied Science Publishers

Rahman SMM (1999) Analysis and prediction of chaotic time series. MSc thesis, IHE Delft, The Netherlands

Solomatine DP (1998) Genetic and other global optimization algorithms - comparison and use in calibration problems In: Babovic, Larsen (eds) Proc. 3rd Int. Conf. on Hydroinformatics, Balkema, Rotterdam, pp 1021-1028

Solomatine DP, Dual KN (2003) Model tree as an alternative to neural networks in rainfall runoff modelling, Hydrological Sciences Journal 48(3):399-411

Solomatine DP, Price RK (2004) Innovative approaches to flood forecasting using data driven and hybrid modelling. In: Liong, Phoon, Babovic (eds) Hydroinformatics 2004: Proc. 6th Int. Conf. on Hydroinformatics, World Scientific Publishing, Singapore, pp 1639-1646

OVERVIEW OF THE NATO SCIENCE FOR PEACE PROJECT ON MANAGEMENT OF TRANSBOUNDARY FLOODS IN THE CRISUL-KÖRÖS RIVER SYSTEM

J. MARSALEK[1]
National Water Research Institute, 867 Lakeshore Rd, Burlington, ON L7R 4A6, Canada

G. STANCALIE
Romanian Meteorological Administration, 97 Soseaua Bucuresti-Ploiesti, 013686 Bucharest, Romania

R. BRAKENRIDGE
Dartmouth Flood Observatory, Dartmouth College, Hanover, NH, USA

M. PUTSAY
Hungarian Meteorological Service, Budapest, Hungary

R. MIC
National Institute of Hydrology and Water Management, Bucharest, Romania

J. SZEKERES
Research Center for Water Resources (VITUKI) Plc, Budapest, Hungary

Abstract. The transboundary (Hungary-Romania) region of the Crisul/Körös river basin suffers from frequent floods, which start in the mountainous headwaters of the basin in Romania and propagate quickly to the plains in Hungary. The NATO Science for Peace Programme has sponsored a project in this region entitled Monitoring of Extreme Flood

---

[1] To whom all correspondence should be addressed. Jiri Marsalek, Environment Canada, National Water Research Institute, 867 Lakeshore Road, Burlington, ON. L7R 4A6, Canada; e-mail: jiri.marsalek@ec.gc.ca

Events in Romania and Hungary Using Earth Observation (EO) Data. NASA also supports this project by providing access to Radarsat and ASTER data via grant funding to one of the co-authors. The main goal of the project is to reduce flood risks in the study area by improved flood management combining hydrological modelling, EO data and GIS facilities; and, refining flood forecasting to increase the forecast accuracy and lead-times. The project is progressing well with an expected completion in 2006.

Keywords: Crisul Rivers, Earth Observation, flood management, floods, Geographical Information System, Körös Rivers, Transboundary flooding

1. Introduction

The transboundary area of the Crisul Alb and Crisul Negru rivers flowing from Romania into Hungary, where they are known as the Körös rivers, is subject to frequent flooding. Such floods originate in mountainous headwaters of the Crisul/Körös river system and propagate to the plains in the lower parts of the basin in Hungary (Marsalek et al. 2004). Examples of recent floods in this transboundary basin include those occurring in June 1974, July-August 1980, March 1981, Dec. 1995-Jan. 1996, March 2000, April 2000 and April 2001. The two spring 2000 floods caused damage of more than $US 20 million on the Romanian territory and the 1980 flood caused damage of more than $US 15 million on the Hungarian territory (Brakenridge et al. 2001).

Historically, there has been a close co-operation between Hungary and Romania in transboundary flood management in this area. Such a co-operation is formalised by the bilateral Agreement for the Settlement of Hydrotechnical Problems, which mentions specifically flood defence and hydrological and meteorological data exchange as the major areas of co-operation. In practical terms, the agreement is implemented by regular meetings of the working groups from the Crisuri Water Authority in Oradea, Romania and Körös Valley District Water Authority (KOVIZIG) in Gyula, Hungary (Brakenridge et al. 2001).

Recognising the need for further improvement of flood management in this transboundary area, an international team was formed, with representatives of Hungary, Romania and USA, and proposed a project on "Monitoring of Extreme Flood Events in Romania and Hungary Using Earth Observation (EO) Data" to the NATO Science for Peace (SfP) Programme (Brakenridge et al. 2001). The proposal was accepted by NATO and the 3-year project started in November 2002. The paper that follows provides an overview of the SfP project and builds on an earlier paper presented at the NATO workshop on Flood Risk Management: Hazards, Vulnerability, Mitigation Measures, which was held in Ostrov u Tise, Czech Republic, in October 2004 (Marsalek et al. 2004).

2. Study area: Crisul/Körös basin

The study area representing the Crisul/Körös transboundary basin was described in detail in the earlier overview paper (Marsalek et al. 2004) and only the most pertinent data are repeated here for completeness. The total study area is about 26,600 km^2, of which 14,900 km^2 lies on the Romanian territory. The Romanian part comprises mountainous areas (38%), hills (20%) and plains (42%), with about 30% of the catchment being forested. On the Hungarian side, the catchment relief represents plains.

The annual precipitation ranges from 600-800 mm/year in lower parts of the basin, and increases to over 1200 mm/year in the mountainous headwaters. The basin frequently experiences large precipitation amounts, with relatively high rainfall intensities, and the frequency of such events seems to be increasing in recent years (Brakenridge et al. 2003).

With respect to runoff generation and transport, there is a great difference between high rates of runoff in headwaters and low rates of runoff in plains. Runoff flood waves form rapidly in the headwaters and move quickly to the plains in the lower part of the basin, which is characterised by relatively slow flows and a potential for inundation (Marsalek et al. 2004).

A hydrometric network is well established in the study area, with 62 hydrometric stations on the Crisul Alb and Negru (and their tributaries); and a number of stations on the Hungarian territory, with those at Gyula and Sarkad being of particular of interest (Brakenridge et al. 2004).

Historically, flood management and particularly flood damage reduction has been practiced in this basin by the respective authorities implementing various structural and non-structural measures. The Romanian area is defended by dykes along the Crisul Alb and Crisul Negru Rivers and such dykes are designed in various river sections for 50 or 100-year return periods. Other structural flood protection measures include permanent retention storage facilities (total volume of 34 x 10^6 m^3) and temporary storage facilities (a total storage volume of almost 80 x 10^6 m^3) (Brakenridge et al. 2003; Brakenridge et al. 2004).

In the Körös valley on the Hungarian territory, low plains are also protected by flood dykes, and following the 1979 flood, construction of emergency detention reservoirs started. Such reservoirs with a storage capacity of 19 x 10^6 m^3 are activated during floods by opening spillways in flood dykes. Detained floodwater inundates areas with low-intensity agricultural activities and causes limited damages (Brakenridge et al. 2003).

3. SfP project objectives

The ultimate goal of the SfP project is to reduce flood risk and damages in the study area by improved flood forecasting and flood defence, through the following measures:

- Providing timely information on the watershed and floods, collected by remote sensing, to a broad range of stakeholders, and thereby improving the efficiency and effectiveness of their action plans for flood defence.
- Creating a database containing EO data, and hydrological and meteorological parameters related to significant flood events, for establishing operational methodologies for detection, mapping and analysis of floods and flooded areas.
- Developing a system for managing remote sensing data related to river flooding, and combining the benefits of GIS, satellite data and hydrological models.
- Providing updated maps of land cover/land use, hydrological networks and thematic maps indicating the extent of the flooded and flood-affected areas.
- Increasing the flood forecast lead time, by improved forecasting, and thereby providing additional time for protecting the human life and property in the study area.
- Providing a basis for restoration and rehabilitation of river courses adversely altered by floods, and also for selecting structural flood protection works.
- Delivering on NATO SfP programmatic criteria, including enhancing co-operation among scientific personnel in the participating countries, training young researchers, disseminating results to the international scientific community, and transferring the tools developed in the study area to another river basin.

Detailed descriptions of the SfP project objectives were presented elsewhere (Marsalek et al. 2004).

4. Progress achieved two-thirds through the project

4.1. IMAGE PROCESSING FOR FLOOD ANALYSIS

EO images are widely used in flood analysis for producing catchment maps; detecting water surface, soil moisture, and inundated areas; and, assisting with remote flow measurement (Muller et al. 1993; Brakenridge et al. 1998). Thus, image processing is important for flood analysis and management. Towards this end, the SfP project team undertook an inventory and documentation of image processing methods, set up an experimental image database, tested and evaluated processing methods, initiated the selection of the best processing methods, and conducted training in image processing (Marsalek 2004; Marsalek et al. 2004).

A database containing EO (Earth Observation), hydrological and meteorological data related to flood events was established to be used to test the processing and analysis algorithms serving to establish operational methodologies for detection, mapping and analysis of flooding. The database contains information concerning the available raw satellite scenes as well as the derived products, and makes it available in a simple format to the SfP project research teams. At the Romanian National Meteorological Administration (NMA) all available raw images were copied on a new Linux file server which can be accessed by the whole remote sensing and GIS group. The satellite image database (SID), which is updated when new images are acquired, was built in Microsoft Works and is available on-line.

4.1.1. Processing MODIS data

MODIS (Moderate Resolution Imaging Spectroradiometer) is a key instrument aboard the Terra (EOS AM) and Aqua (EOS PM) satellites. Terra MODIS and Aqua MODIS view the entire Earth's surface every 1 to 2 days, acquiring data in 36 spectral bands, or in groups of wavelengths. The data available from MODIS are highly suitable for flood warning and management, because they are available in real time (or nearly real time), can be rapidly processed and disseminated, cover a wide area, and are abundant and inexpensive (Brakenridge et al. 2003; Marsalek et al. 2004).

The choice of the most efficient method for water identification and mapping depends on the type of satellite information (optical or radar) as well as on the spatial resolution this information could provide. From the operational point of view, during dangerous floods, preference is given to simple and fast methods. Towards this end, a MODIS processing algorithm was implemented in the RS and GIS team group of the NMA in Bucharest. This algorithm allows extracting images of water bodies from the MODIS data, MOD02QKM (250 m resolution) and MOD02HKM (500 m resolution). The main steps for data processing are: (a) removal of the bow-tie effect from the MODIS image, (b) importing the two types of MODIS data (HKM and QKM) into the ERDAS Imagine environment, (c) re-sampling the HKM image to a 250 m resolution, (d) resolution merging of HKM and QKM data to obtain an image with three bands at a 250 m resolution, (e) MODIS image geo-referencing using an ASTER image, (f) performing the supervised classification in four classes, and, (g) creating a binary mask by merging the three classes which differ from water.

The Hungarian Meteorological Service (HMS) has developed two methods for creating the water mask using multi-spectral MODIS images (Putsay 2003; Putsay 2004). The first one is an automated method, which is quick, but may mis-detect some shadow pixels and not detect some water pixels. Using reflectance (R) signals in the 0.87 and 1.6 µm channels, optimum thresholds for separating water from shadows were determined. The second method, referred to as a quasi-automated method, is more accurate than the former one, but needs an interactive correction by removing the mis-detected cloud shadow vectors manually. The quasi-automated method was developed by modifying the water mask equation and its thresholds. Subsequently, the method was tested on MODIS images of the spring 2000 flood on the Körös rivers, using images with proper solar zenith, and bow-tie and geo corrections. After preparing the water mask raster image, it was converted to vectors, and mis-detected cloud shadow vectors were removed manually. The new equations and thresholds, using both 0.87 and 1.6 µm channel reflectance data, were defined (Putsay 2003; Putsay 2004).

In applications of this method to the 2000 flood, water masks derived for six days from Apr. 9 to May 18, 2000, were visualised to demonstrate the variation of the inundated area. Water masks were then overlaid on the land use map (derived from an ASTER image) to demonstrate the extent of inundation.

4.1.2. *Use of satellite radar images in water extent mapping*

Radar images are more difficult to interpret than optical images, but are not affected by clouds and can be therefore useful for flood mapping. Consequently, some radar processing techniques were also investigated to identify the most efficient one for flood inundation mapping. The satellite data used were RADARSAT-1 Standard Beam, Full Resolution images from 1999, 2001 and 2002, and served to obtain a mosaic for the whole Romanian study area in the Crisul Alb and Crisul Negru basins. The image processing was done using the dedicated software ERDAS Imagine, version 8.6, ENVI version 4.0, as well as ArcGIS version 8.0. The main processing steps included: (a) data import, (b) geo-correction and re-projection in

UTM / WGS 84, Zone 34, (c) "speckle" effect removal using various radar texture filters (Mean, Median, Lee-Sigma, Local Region, Lee, Frost, Gamma MAP) with different window sizes (3x3, 5x5, and 7x7), (d) supervised and unsupervised classification procedures to identify water, and (e) creating the water mask. The best water mask was obtained using the Gamma MAP filtered image (5x5 size).

4.2. DEVELOPMENT OF A DEDICATED SUBSYSTEM (DSS) BASED ON REMOTE SENSING (RS) AND GIS

This subsystem serves to acquire, analyse and interpret data; manage data; handle and prepare data for rapid access; update information; restore data, and prepare value-added information (Putsay 2004). The DSS system is based on both remote sensing and GIS, and its central platform (DSS server) was set up at the Romanian NMA, where satellite data are received and processed, and further distributed. DSS also allows data exchange among the Project partners.

The Romanian team proceeded with developing a GIS for the entire (Romanian) study area using cartographic maps (1:100,000) and updating such information by recent satellite images (e.g., the hydrographic network, land cover/land use) and field measurements (e.g., dyke and canal network) (Brakenridge et al. 2004). Towards that end, inventories of all meteorological and hydrometric stations, hydrotechnical structures (including dykes, drainage works, agricultural drainage, irrigation systems, flood retention reservoirs, and polders) in the study area have been completed, and cross-sectional profiles of main river beds have been measured in the field (Marsalek et al. 2004).

The structure of the GIS database, represented by the spatial geo-referenced information ensemble (satellite images, thematic maps and series of meteorological and hydrological parameters, and other exogenous data), was planned to serve the study objectives with respect to evaluation and management of information related to flood occurrence and development, and the assessment of damages inflicted by floods. The final structure of the GIS database contains the following info-layers:

- administrative limits (county and country borders);
- sub-basin and basin limits;
- land topography (90-metre DEM);
- hydrographic network, dykes and canals network;
- transportation network (roads, railways);
- municipalities;

- meteorological station network, rain-gauge network, hydrometric station network; and,
- land cover/land use, updated from satellite images.

The databases developed for the Romanian and Hungarian study areas have been re-projected to the UTM (Universal Transverse Mercator), zone 34 (datum WGS84) coordinate system.

For the most flood vulnerable area, situated in the plain of the Crisul Alb/Negru, demarked at its Eastern boundary by the Ineu-Talpos and at its northern boundary by the Crisul Repede basin, a more precise GIS database is being developed using information derived from 1:5,000 and 1:10,000 topographic plans and supplemented by information obtained from field surveys and satellite images. The structure of the GIS database for this area is similar to the main database, but it contains more detailed information for features like dykes, canals and river networks.

The Hungarian team collected the digital maps of their study area, which were derived from 1:100,000 scale topographic maps, and the ARCVIEW shape files, and shared this information with the Romanian Team. Connections of objects crossing the Hungary-Romania border (e.g., roads, dykes, rivers, canals, etc.) were verified and corrected at a working meeting of both parties. The land cover/land use information in the topographic maps was derived from the 1:100,000 scale CORIN database, which was developed from satellite images collected from 1990 to 1992. Consequently, such information was updated using (vectorised) ASTER images prepared in 2002-03 by HMS. These maps will be further used in flood risk mapping.

4.3. INTEGRATED METHODS FOR FLOOD MANAGEMENT

In this task, a flood database was established for the study area (from the Romanian and Hungarian national databases) and validated, maximum discharges of various return periods were calculated, synthetic flood hydrographs were developed, and land cover/land use maps are under preparation.

4.3.1. *DEM Construction and Integration into the GIS Database*

To develop a digital elevation model (DEM) for the study area, the shape and elevation information, extracted from individual maps, were merged, corrected and interpolated (Stancalie et al. 2003; Stancalie 2004). The main challenge in this task was the large volume of data to be considered in interpolations. Among the different interpolation methods tested for this area, the Triangulated Irregular Network (TIN) method was found the fastest but its output did not meet all expectations, mainly because of the low energy of the relief, with elevations varying between 86 and 113 metres above the sea level. The best results were obtained with the Kriging interpolation method, which however was very demanding with respect to the computing times. Consequently, the input data had to be divided into smaller datasets interpolated separately. The results were merged into a five-metre cell size grid with a sub-metric vertical accuracy. The final task was to procedurally integrate dykes, canals and roads embankments into the digital elevation model. The height or depth of these structures was obtained from field surveys or conventional maps. The Hungarian team contributions were described in Section 4.2, and were merged with the Romanian data in late 2004. Thus, a single DEM for the study area has been developed.

4.3.2. *Characteristics of extreme floods*

The characteristics of extreme floods, i.e. peak flows, volumes and durations, and their probabilistic distributions, are needed and were determined by the Flow (Q)-Duration (d)-Frequency (F) method. The estimates of low-frequency flood quantiles were produced by the GRADEX method, in which maximum rainfall distributions are used to extrapolate hydrometric data (Marsalek 2004; Brakenridge et al. 2004). For each flood event, characteristic flows were determined, partial-duration series of these variables were fitted by the exponential law, and extrapolated to lower frequencies by gradually replacing the flow distribution slope by the rainfall gradex, as reported earlier (Marsalek et al. 2004).

4.4. METHODOLOGIES FOR FLOOD FORECASTING

In preparation for hydrological modelling of the study area, metadata files were prepared as described below.

4.4.1. *Metadata file*

Metadata are the definitional data that provide information about, or documentation of, other data managed within an application or environment. For example, metadata would document information about: (a) data elements or attributes, (name, size, data type, etc.), (b) records or data structures (length, fields, columns, etc.), and (c) data themselves (where they are located, how are they associated, ownership, etc.). Metadata may include descriptive information about the context, quality and condition, or characteristics of the data.

The use of metadata files to describe the GIS database for the study area in the Crisul/Körös basins provides a common set of terminology and definitions of geospatial data for the following topics:

- Identification information: basic information about the data set (title, geographic area covered, rules for acquiring or using the data).
- Data quality information: assessment of the quality of the data set (positional and attribute accuracy, completeness, consistency, sources of information, and methods used to produce the data).
- Spatial data organisation information: the mechanism used to represent spatial information in the data set (the method used to represent spatial positions directly, such as raster or vector, and indirectly, such as basin or county codes, and the number of spatial objects in the data set).
- Entity and attribute information: information about the content of the data set, including the entity types and their attributes and the domains from which attribute values may be assigned (names and definitions of features, attributes, and attribute values.
- Distribution information: information about obtaining the data set (location of the data, and available formats).

4.4.2. Hydrological modelling

Two types of hydrological models are used in the project – a flood forecasting model VIDRA on the Romanian territory and the output hydrographs from this model will be routed in the Hungarian part of the basin by the HEC-RAS model of the U.S. Army Corps of Engineers (HEC-RAS: Hydrologic Engineering Center - River Analysis System). The VIDRA model (Brakenridge et al. 2004) simulates the rainfall-runoff processes taking place in a watershed by conducting the following computations (Marsalek et al. 2004):

- Sub-basin snowmelt estimation, using the degree-day method;
- Computation of the average rainfall in each sub-basin, by weighting the rainfall and snowmelt data measured in the meteorological network;
- Calculation of the effective rainfall over each sub-basin by subtraction of infiltration and evapotranspiration abstractions from the average water inflow, using the deterministic reservoir model PNET;
- Integration of the effective rainfall on hill slopes and in the primary river network, which results in runoff hydrograph formation in each sub-basin, using the instantaneous unit hydrograph as a transfer function of the hydrographical system;
- Superimposition of the flood waves formed in each sub-basin and their routing along the main river channel using a non-linear model based on the analytical solution of the Muskingum method; and,

- Flood wave attenuation through reservoirs, using a reservoir co-ordinated operation method.

The hydrographs, computed by the VIDRA model, will be inputted into the HEC-RAS model, which can model steady and unsteady hydraulic/hydrologic phenomena, taking into consideration a wide range of options, including the following:

- compound river bed cross-sections (main channel and floodplain),
- bridges, sluice weirs, side weirs, and other structures,
- reservoirs,
- horizontally and vertically changing roughness coefficients, and
- morphological effects (braiding, joining and diffusing channels, etc.).

The input time series comprises discharges at the upstream end and stages at the downstream end of the river system. Computed results may be demonstrated in tabulated, graphical and animated forms. The animation form is particularly effective for introducing complicated physical processes to the decision makers and the concerned public.

So far, the geometry of the river bed and the hydraulic parameters have already been entered into the model, and the model is being calibrated and tested. In the next phase, different flood scenarios will be modelled and evaluated for assessing the most critical flood situations.

4.4.3. *Calibration of the VIDRA model for the Crisul Alb and Crisul Negru River basins for major rainfall – runoff events*

The calibration process for the VIDRA model conducted in the SfP project includes the following steps: (a) Basin discretisation into homogenous units (sub-basins) and the river network into homogenous reaches (sectors); (b) estimation of the initial model parameters using the parameter generalisation relationships; and, (c) the overall model parameters calibration using the historical database with major rainfall–runoff events.

The overall calibration of the VIDRA model parameters has been achieved by comparing the modelled values with the measured ones for the following variables: discharge hydrographs, the maximum discharge, Q_{max} and the time of its occurrence (at the hydrometric stations); the flood volume W_s; the depth of runoff h_s; and, the runoff coefficient, α, defined as the ratio between the depths of runoff and precipitation for each sub-basin.

The criteria of agreement between the modelled and measured values were chosen as the relative error, the Nash coefficient and the time interval between Q_{max} observed and Q_{max} simulated. One such a comparison between modelled and measured hydrographs for the Crisul Alb River at the Chisineu Cris hydrometric station is shown in Fig. 1.

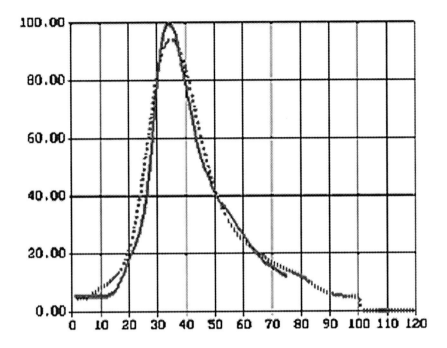

Fig. 1. The recorded (full line) and simulated (dotted line) hydrographs for the Crisul Alb River, at the Cisineu Cris gauging station (The vertical axis shows the percentage of the maximum discharge of 258 m^3/s, the horizontal axis shows a number of computational time steps $\Delta t = 3$ h)

4.4.4. *Assessment of the parameters derived from images and the GIS database*

Four parameters used in hydrologic modelling for individual sub-basins are derived from satellite images and the GIS database: mean elevation, mean slope, soil texture and the forest cover coefficient. For gauged stations, the information from the GIS database and the hydrological database is merged and used to derive generalised relationships, which are then used, in conjunction with the GIS database, to estimate the initial model parameters for the ungauged sites.

4.5. NEW SATELLITE-BASED APPLICATIONS AND PRODUCTS

4.5.1. *Producing updated land cover/land use maps*

Various supervised and unsupervised classification procedures have been tested on ASTER data to produce land cover/land use maps. The resulting classes were finally grouped into eight classes: forests, pastures, winter crops, summer crops, municipalities, orchards, bare soils and water. The municipalities' images were digitised as an info-layer and overlapped onto the classification. For the Crisul Alb and Negru basins and the Hungarian study area, the classified images were merged and mapped in UTM projection, WGS 84, Zone 34 N.

4.6. PROGRAMMATIC OBJECTIVE ACHIEVEMENTS

So far, good progress has been achieved towards meeting the programmatic objectives. Project visibility is maintained through presentations at various international meetings and the project web site. There are about 15 young scientists working on various project tasks at the five agencies involved. Also, staff members have been trained at the Dartmouth Flood Observatory (USA), DHI (Denmark), and elsewhere. Deliveries on other objectives are being planned.

5. Concluding observations

The NATO Science for Peace Programme is an important mechanism for supporting scientific research in the transition countries of Central and Eastern Europe. The project on floods in the transboundary region of Hungary-Romania fits well the objectives of the SfP Programme, with respect to its focus on research with high societal value (protecting human life and property), conduct of leading-edge research in terms of Earth Observations and their use in flood management, enhancing collaboration between the two neighbouring countries and others, and fostering collaboration among international experts. The progress achieved so far holds promise of successful completion of this SfP project.

References

Brakenridge RG, Tracy BT, Knox JC (1998) Orbital SAR remote sensing of a river flood wave. Int. J. Remote Sensing 19(7):1439-1445

Brakenridge RG, Stancalie G, Ungureanu V, Diamandi A, Streng O, Barbos A, Lucaciu M, Kerenyi J, Szekeres J (2001) Monitoring of extreme flood events in Romania and Hungary using EO data (project plan). September 2001, Bucharest, Romania

Brakenridge RG, Stancalie G, Ungureanu V, Diamandi A, Streng O, Barbos A, Lucaciu M, Kerenyi J, Szekeres J (2003) Monitoring of extreme flood events in Romania and Hungary using EO data (progress report). May, 2003, Hanover, NH, USA

Brakenridge RG, Stancalie G, Ungureanu V, Diamandi A, Streng O, Barbos A, Lucaciu M, Kerenyi J, Szekeres J (2004) Monitoring of extreme flood events in Romania and Hungary using EO data (progress report). May, 2004, Hanover, NH, USA

Marsalek J, Stancalie G, Brakenridge R, Ungureanu V, Kerenyi J, Szekeres J (2004) NATO Science for Peace project on management of transboundary floods in the Crisul- Körös river system (Romania-Hungary). In: Flood Risk Management: Hazards, Vulnerability, Mitigation Measures, Proc. of NATO Advanced Research Workshop, Ostrov u Tise, Czech Republic, Oct. 6-10, pp 191-202

Marsalek J (1993) Monitoring of extreme flood events in Romania and Hungary using EO data - Technical Progress Report No. 3, NWRI, Burlington, Ontario, Canada

Muller E, Decamps H, Dobson MK (1993) Contribution of space remote sensing to river studies. Freshwater Biology 29:301-312

Putsay M (2004) Creating a water mask using a threshold technique on multispectral MODIS images http://nato.inmh.ro/Media/nato2_presentations.html

Putsay M (2003) Creating a water mask using a threshold technique on multispectral MODIS images, Report of visit of Maria Putsay to Dartmouth Flood Observatory, November 25

Stancalie G (2004) Contribution of Earth Observation Data to flood risk analysis. In: 22nd Conference of the Danubian Countries on the Hydrological Forecasting and Hydrological Bases of Water Management, Brno, Czech Republic, Aug. 30-Sep. 3 (CD-ROM)

Stancalie G, Diamandi A, Ungureanu V, Stanescu VA (2003) Sub-system based on remote sensing and GIS technology for flood and related effects management in the framework of the NATO SfP "TIGRU" project. In: First Annual Session of the NIHWM, Bucharest, Sep. 22-25, (CD-ROM)

COPING WITH UNCERTAINTIES IN FLOOD MANAGEMENT

ISTVAN BOGARDI [1]
Department of Civil Engineering, University of Nebraska-Lincoln,
Lincoln, NE, 68588, USA

Abstract. Main elements of risk-based design of flood management are considered: flood as a load or exposure, resistance/capacity of flood control works, and consequences of possible flooding (economic, human life, ecological). All the three elements can be described with much uncertainty stemming from limited information and imprecise models/definitions. As a consequence, the classical risk-based design is often incorrect and cannot be defended, because it may lead to over or under design; both resulting in economic, social and ecological losses as well as unwarranted human life hazards. Statistical/probabilistic methods have difficulties to encode such uncertainties. Fuzzy logic, as an alternative, is shown to be possibly better applicable to the present situation. It is shown how that methodology can be used to find the best flood management alternative under uncertainties in view of satisfying – the often conflicting – economic, human life and ecological objectives.

Keywords: risk-based flood management, bootstrapping, fuzzy logic, multi-criterion decision making

1. Introduction

The purpose of the paper is to present the classical probabilistic and fuzzy logic formulations of risk-based design of flood management with application to a flood risk management case. It is common to distinguish four main elements of risk analysis (Haimes 1998). These are: (a) Exposure, H, which may be represented as a natural hazard (e.g., flood stage), (b) Capacity/Resistance, *R* (e.g., strength of a dyke, or capacity of a reservoir), (c) Failure event, $H > R$, and (d) Consequences of the failure (e.g., economic, human life, ecological, etc.). Risk analysis is necessary in cases where uncertainties are inherent in any of the four elements. The traditional probabilistic formulation considers these uncertainties as random variables, represented for instance by the respective probability density functions: *g(H)* and *f(R)*. Then the probability of failure, $P(H>R)$ can be calculated from *g(H)* and *f(R)* The economic, human life and ecological consequences of failure can be represented by the respective loss functions *L(H,R)*.

2. Probabilistic formulation of flood management

The general objective of flood management includes three main elements: (a) minimising economic losses ($): property, production, transportation, etc., (b) minimising casualties, and (c) minimising ecological losses. These three inherent elements are always present to various degrees in actual flood management design. Consequently, the "best" design option should be obtained by a trade-off analysis, using multi-criterion decision making tools, as shown in the application section of this paper. In practice, however, mostly the economic objective is dominating and the other two may be considered as constraints. In the following, the probabilistic formulation will be specified with the help of the economic objective.

The probabilistic formulation of risk-based design is based on the statistically expected consequences and aims at selecting the best design by minimising the expected consequences. Specifically, the engineering risk *ER* corresponds to the sum of expected annual economic losses and the annual cost of flood management:

$$ER = \int_{R}^{\infty} L(H)\,g(H)\,dH + \text{cos}\,t(R) \qquad (1)$$

If the resistance/capacity, *R*, of the flood management scheme is assumed to be precisely known that is deterministic, the above integral may be solved approximately by numerical integration (Figure 1).

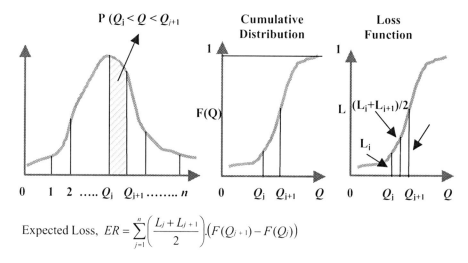

Expected Loss, $ER = \sum_{j=1}^{n}\left(\dfrac{L_j + L_{j+1}}{2}\right).(F(Q_{j+1}) - F(Q_j))$

Fig. 1. Calculation of expected annual flood losses (resistance is deterministic)

The economically optimal flood design corresponds to the minimum of the engineering risk. However, the optimal design may be different if the other two main objectives (human life protection and ecological preservation) are also considered (as in the application example).

This probabilistic formulation has several known shortcomings, including:
1. Calculation of the expected risk becomes more difficult if the resistance – realistically – is imprecisely known;
2. Management of the low-failure-probability/high-consequence case may be misrepresented by the expected value as shown in Fig. 1;
3. Selection of the probability models (flood probabilities and especially resistance probabilities) are often uncertain or even arbitrary, while the results may be quite sensitive to this choice;
4. Statistically meaningful data on flood exposure and/or resistance are often lacking;
5. The consequence functions, for instance flood losses in the domain of extreme floods, are commonly quite uncertain (Figure 2).
6. The covariances (dependencies) between the various types of exposures and resistances are commonly unknown. Again, the results are highly sensitive in this respect.

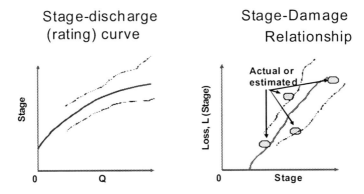

Fig. 2. Uncertainties in the relationship between stage and discharge, and flood storage and flood losses

There are various methods to deal with some of these shortcomings. The conditional expected value formulation (Haimes 1998) helps define the low-probability/high-risk dilemma in a more appropriate way. The probability bounds analysis (Ferson et al. 2002) accounts for the uncertainty of the probability model selection and the unknown covariances. A Bayesian formulation may be used if data are unavailable for frequency interpretation (e.g., Carlin and Louis 2000).

Listed as a major uncertainty (Figure 3), the effect of limited statistical information on probability estimates (e.g. 1% frequency flood) can be evaluated with so-called parametric and non-parametric approaches.

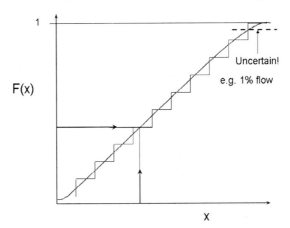

Fig. 3. Uncertainty of probability estimate

The parametric approach assumes the knowledge of the distribution type, F (e.g. normal), and the distribution of its parameters (e.g. mean and variance). The main steps of the parametric approach include:

1. Generate a random realisation of parameters, α_1 (using the inverse of the parameter distributions independently with random numbers between 0-1), using the "known type" of distribution F, solve $X_1 = F^{-1}(\alpha_1, p); p$ – e.g. 1%
2. Repeat 1 and 2 n-times (e.g. 100 times) to obtain: $X_1,..,X_n$, and perform frequency analysis of $X_1,..,X_n$ to obtain confidence limits of the quantile (e.g. p=1% flood)

The main advantage of the non-parametric approach is that neither the type of the underlying distribution, nor its parameter distributions need to be specified. One of the common non-parametric approaches, the bootstrap method (Chernick 1999), uses observation data: $x_1,...,x_n$ in the following steps (Figure 4):

1. Construct empirical cumulative distribution
2. Generate from the empirical distribution a new random sample $y_1,..,y_n$ using $u_1,..,u_n$ random numbers between 0-1: $y_1 = F_x^{-1}(u_1),..,y_n$
3. Construct frequency distribution of the "new" bootstrap sample, y
4. Repeat 2 and 3 many times (the maximum number of bootstrap samples is 2^n)
5. Obtain from each sample the selected p-quantile value
6. Construct frequency distribution of p-quantile values, $f(x \mid p)$, and determine, say, the 80% (90-10) confidence limits.

The next section will show that non-probabilistic methods, namely the fuzzy set formulation, can be considered as a practical alternative to account for the uncertainties facing flood managers. It is applicable when all of the above listed six shortcomings are dealt with simultaneously.

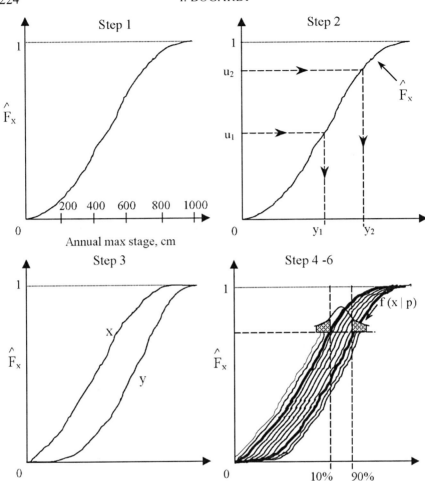

Fig. 4. Main steps of the bootstrap method

3. Fuzzy logic formulation of risk-based design

Fuzzy logic is an alternative to statistical methods for defining uncertainties. In that respect, fuzziness represents situations where membership in sets cannot be defined on a yes/no basis because the boundaries of the sets are vague. The central concept of the fuzzy-set theory is the membership function, which represents numerically the degree to which an element belongs to a set. As the degree increases, the value of the membership function for the element also increases.

In a classical set, a sharp or unambiguous distinction exists between the members and non-members of the set. In other words, the value of the membership function of each element in the classical set is either *1* for members (those that certainly belong to the set) or *0* for non-members (those that certainly do not). However, it is sometimes difficult to make a sharp or precise distinction between the members and non-members of a set. For example, the boundaries of the sets of "highly contaminated water", "deep wells", or "high flood protection capacity" are fuzzy.

Since the transition from member to non-member appears gradual rather than abrupt, the fuzzy set introduces vagueness (with the aim of reducing complexity) by eliminating the sharp boundary that divides members of the set from non-members (Klir and Yuan 1995). Thus, if an element is a member of a fuzzy set to some degree, the value of the function can be bounded, say between 0 and 1. When the membership function of an element only can have values 0 or 1, the fuzzy-set theory reverts to the classical-set theory. The membership value of a real number reflects the "likeliness" of the occurrence of that number; the level sets (intervals in this case) reflect different sets of numbers with a given minimum likeliness (Zimmermann 1991). Any real number can be regarded as a fuzzy number and often is called a *crisp* number in fuzzy mathematics. The simplest type of fuzzy number is triangular, that is, linear on either side of the peak. Figure 5 gives an example of a triangular fuzzy number (TFN). The fuzzy number displayed in Figure 5 can be described by the values of z at points $a_1 = 9$, $a_2 = 18$ and $a_3 = 22$. Thus, $A = (a_1, a_2, a_3)$ completely characterises that fuzzy number.

Since the early application of fuzzy logic to water resources (Bogardi et al. 1983), there has been a great deal of research, and at present, fuzzy logic has become a practical tool in water resources analysis and decision-making. See a review in Bogardi and Duckstein (2003).

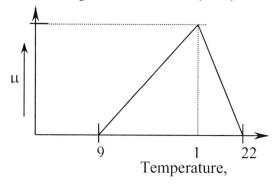

Fig. 5. Membership function of temperature: "higher than 9, lower than 22, around $18°$"

The main idea of using fuzzy sets in flood management design is quite simple. Namely, if uncertainty in any element of the risk is described as a fuzzy set (e.g. a fuzzy number instead of a fixed value), the corresponding risk can be also calculated as a fuzzy set. There are numerous possibilities specific to flood management.

For instance, the fuzzy engineering flood risk may consider a probabilistic flood exposure H, say peak water level (stage), an uncertain capacity of flood protection, R (expressed as a fuzzy number), and an imprecise loss function, $L(H,R)$ as illustrated in Figure 2. Then the engineering risk $\widehat{ER}$ will be obtained as a fuzzy number:

$$\widehat{ER} = \int_R^\infty \widehat{D}(H)g(H)d(H) \tag{2}$$

The numerical calculation of the fuzzy risk (i.e., calculation of a function of fuzzy numbers) is a relatively simple task using so-called fuzzy arithmetic, that is, the operation with fuzzy numbers. Fuzzy arithmetic can be performed by several tools, such as the vertex method (e.g., Dong and Shah 1987).

Often, a mixed probabilistic/fuzzy-set formulation is the most appropriate way to encode uncertainties. The next section provides such an example; it will also indicate that the fuzzy-risk formulation is amenable to risk management purposes.

4. Flood management example

The classical risk-based approach will be improved by the use of fuzzy logic. It is shown how that methodology can be used to find the best flood management alternative under uncertainties in view of satisfying – the sometimes conflicting – human life and ecological objectives.

A reach of the Santa Cruz River in southern Arizona is used to illustrate an application of the mixed probabilistic/fuzzy-set risk management approach. This simplified case consists of a binary design problem: to provide protection or not against the maximum floods.

The four main elements of this flood risk analysis include:

1. Flood hazard as the exposure: here to the flood discharge H,
2. Resistance, or capacity of flood control works R,
3. Failure event, whenever $H > R$, and
4. Consequences of the binary decision - here, the economic losses of flooding, the cost of protection, and the ecological impacts of human intervention (protection) are considered.

The problem is formulated using:

(a) Two states of nature: flooding S_1, when $H \geq R$ and S_2: $H < R$,

(b) A Bernoulli probability model of state occurrences, i = 1, 2:

$P(S_1) = P(L \geq R) = p$

$P(S_2) = P(L < R) = 1 - p$

(c) Two actions: a_1: protect and a_2: do not protect

(d) Annual flood losses, K and protection costs, C.

The following uncertainties – quite common in flood management design - are represented by triangular fuzzy numbers:

1. Exceedance probability: $p = (7, 8, 10) \times 10^{-4}$
2. Annual economic losses: $K = (4, 5, 6) \times 10^6$
3. Annual protection costs: $C = (3, 4, 5) \times 10^3$, a cost of "about" 4000.

The economic consequences of the two actions are shown in the following economic loss matrix:

I	S_i	A_1 protect	A_2 do not	$P(S_1)$
1	H > R	C	K	p
2	H ≤ R	C	0	1-p

The expected annual economic losses D, corresponding to actions a_1 and a_2, are:

$$D(a_1) = C(.) p (+) C(.)(1-p) = C \quad (3)$$

$$D(a_2) = K(.) p \quad (4)$$

Fuzzy arithmetic is used to calculate $D(a_1)$ and $D(a_2)$ as fuzzy numbers to obtain:

$$D(a_1) = (3, 4, 5) \times 10^3 \text{ and } D(a_2) = (2.8, 4, 6) \times 10^3 \quad (5)$$

To compare $D(a_1)$ and $D(a_2)$, the fuzzy mean, that is, the centre of gravity of the fuzzy numbers is used:

$$D(a_1) = 4 < D(a_2) = 4.27 \quad (6)$$

Thus, on a purely economic basis, action $a = a_1$, *protect*, is preferred to action a_2.

The ecological effect of flood protection is measured by the number of species lost if a flood-control structure is built. The initial number of the species is assumed to be known precisely: $N_1 = 60$. Diversity after construction, which is not predictable precisely, is represented by a fuzzy number: $N_2 = N(a_2) = (10, 25, 40)$, corresponding to about 25 species left, but not less than 10 and not more than 40.

The ecological consequences of the two actions are shown in the following ecological matrix:

I	S_i	a_1 protect	A_2 do not	$p(S_1)$
1	H > R	N_2	N_1	P
2	H ≤ R	N_2	N_1	1-p

To evaluate the two actions a_1 and a_2 from a joint economic/ecological aspect, a fuzzy trade-off relationship is constructed (Figure 6).

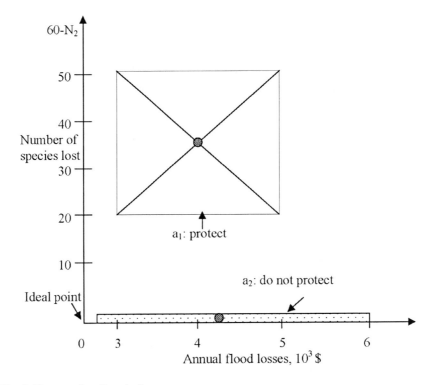

Fig. 6. Fuzzy trade-off analysis

Action a_1 *(protect)* is represented by a two-dimensional fuzzy number reflecting an uncertain economic and ecological quantity. Action a_2 *(do not protect)* does not show ecological uncertainty and is thus represented by a one-dimensional fuzzy number projected at the horizontal segment

$(2.8, 4, 6)$. The action closest to the ideal point - neither economic nor ecological losses - is selected: the distance between two fuzzy numbers (here, one point and one fuzzy number) thus must be defined. Taking again the mean value representation for a fuzzy number, it is found that action a_2 is preferred to action a_1. Thus, taking ecological effect into account leads to a reversal of the decision that would be taken on purely economic grounds.

5. Conclusions

1. The main elements of risk-based design - flood exposure, protection capacity and the consequences - are commonly uncertain or rather imprecisely known.
2. The probabilistic formulation of imprecision/uncertainty may have several difficulties especially when statistical data are limited or even unavailable.
3. Neglecting the uncertainties or their incorrect representation could lead to false flood protection design.
4. Fuzzy-logic formulation is an alternative to the probabilistic formulation.
5. The "best" design is a trade-off among the often competing objectives - economic, human life protection and ecological - of flood management.
6. A simplified flood-risk management illustrates a case where:
- Probabilistic and fuzzy uncertainties are involved in exposure, resistance and consequences,
- Economic and ecological consequences are considered, and
- A trade-off analysis can identify the preferable action.

References

Bogardi I, Duckstein L, Bardossy A (1983) Regional management of an aquifer under fuzzy environmental objectives. Water Resources Research 19(6):1394-1402

Bogardi I, Duckstein L (2003) The fuzzy logic paradigm of risk analysis. Risk-based decision making in water resources X, ASCE Press, Reston, VA

Chernick MR (1999) Bootstrap Methods: a Practitioner's Guide. New York, Wiley

Carlin BP, Louis TA (2000) Bayes and Empirical Bayes Methods for data analysis. Chapman and Hall

Dong W, Shah HC (1987) Vertex method for computing functions of fuzzy variables. Fuzzy Sets and Systems 24:65-78

Ferson SL, Ginzburg J, Hajagos V, Tucker WT (2002) Terrorism's implications for uncertainty calculi. Risk-based decision making in water resources X, ASCE Press, Reston, VA

Haimes YY (1998) Risk modeling, assessment, and management. New York, Wiley

Klir GJ, Yuan K (1995) Fuzzy sets and fuzzy logic: Theory and application. Prentice Hall, New York

Zimmermann HJ (1991) Fuzzy set theory and its applications, Boston, Kluwer-Nijhoff

TRANSBOUNDARY FLOODS IN AZERBAIJAN

RAJAB MAMMEDOV[1]
*Association of International Hydrological Programme, Baku,
AZ1000, Azerbaijan*

Abstract. An analysis of transboundary floods on the whole territory of the republic of Azerbaijan is presented. Average characteristics of spring floods and maximum streamflows of the Azerbaijan Rivers are given.

Keywords: spring floods, autumn high waters, landslides, flood storage reservoirs, flood protection measures.

1. Introduction

Every year floods in Azerbaijan cause significant damage (Rustamov and Gashgay 1989). Thus, flooding is a yearly occurring phenomenon especially on the Kura River. It may be induced by rainfall or extreme snowmelt.

A major problem is the flooding of agricultural land and villages in the countryside. Due to deforestation little water is stored in the hills and mountainous areas. As a result the water is running off faster and floods downstream areas.

Another problem caused by deforestation and irrigation is the greatly increased sediment load in rivers during the last 30 years. Part of the sediment is deposited in dam reservoirs, but major sedimentation occurs in the lower section of the Kura in Azerbaijan (Tacis Project 2003).

The sedimentation problems need to be seen in conjunction with erosion in main river tributaries, which may result in blockage of smaller tributaries with sediment and reduced effective use of flood plains of these tributaries. Flood protection is difficult to achieve at the points of impact, especially

---

[1] To whom all correspondence should be addressed. Rajab Mammedov, Association of International Hydrological Programme 4, Hasan Aliyev Street, 190 Baku, AZ1000 Azerbaijan; e-mail: rajab@kura-aras.org

where meandering river flood plains and wetlands are lacking. Where such buffers for floodwater are not available, solutions should be sought in the upstream sections to reduce flood inundation and compensate for catchment deforestation. Another problem is due to the fact that major waste disposal sites are located near the rivers. Flooding may result in transport of waste into the river, even though this has not happened as yet in most places. During the spring period, river floods damage the infrastructure and private properties, and sometimes cause losses of human life. In the Kura River basin (Tacis Project 2003), floods are generally caused by heavy rainfall and in the Aras River basin, the main cause is snowmelt. In particular, when suddenly occurring high temperatures and relatively warm rain trigger high runoff, floods can be devastating and destroy bridges, riverbank protection, cultivated fields and whatever is in the way of floods.

In parts of the country with steep slopes and loose weathered surface materials, high rainfall and related runoff often cause landslides, which tend to be more damaging than runoff with low concentrations of solids (Museyibov 1998). Complex measures are needed for the reduction of damages caused by transboundary floods.

2. Analysis of the current situation

The basic part of water resources of Azerbaijan comprises transboundary rivers: the Kura and Aras in Southern Caucasus (Mamedov 2002). In addition to the main river channels, there are 9 tributaries of the Kura and 9 tributaries of the Aras, which are also transboundary rivers. The drainage area of the Kura, including the drainage area of the river Aras, is 188,000 km^2. The general length of the Kura River is 1,515 kms; 174 km on the territory of Turkey, 592 km in Georgia and 749 km in Azerbaijan (Rustamov and Gashgay 1989; Verdiyev 2002; Museyibov 1998).

The general drainage area of the transboundary Kura is 30,200 km^2, of which 7,400 km^2 is on the territory of Azerbaijan. The transboundary tributaries of the Kura are the Hrami, Qanihchay (Alazani), Qabirrichay (Iori), Indjasu, Dashsalaxli, Akstafachay, Qasansu, Ahindjachay, and Tauzchay.

High spring waters are typical for the rivers of the Kura basin. From 40 to 80% of the annual streamflow volume is conveyed by these rivers during the periods of high water. By comparison, conveyance of a greater share of the annual streamflow (up to 55-65%) by the spring high water is characteristic for the Qirdimanchay, Ganjachay, Kurakchay, Terter and some other rivers (Table 1, 2).

For example, in the region of Graqkesaman and Kirzan, floods are observed 3-5 times per year and during high water. Flood plains, 15-20 m wide, fill with water 0.8-1.2 m deep, during the periods of flooding lasting 5-15 days. Once in 5-10 years, during the passage of high flows, continuous flooding is observed.

During high water, relatively small flows are observed on the rivers with strong flow control by groundwater, such as the Zabuhchay. On the majority of rivers the high water begins in March. The maximum annual discharge is typically observed during the spring high water. The maximum discharge of the majority of rivers is formed by rainwater and snowmelt (Table 2).

The duration of spring high water for the majority of the rivers is on average about 100 days. Late ending high water is typical for the Ganjachay and Terterchay rivers.

The autumn floods occur on the rivers, which flow down the southern slope of the Major Caucasus (Mamedov 1989). The maximum discharges of the rivers of the southern slope of the Big Caucasus during the spring-summer period are always formed by combined rainfall and snowmelt. Thus the role of rainwater in flood formation is prevailing.

The analysis of the hydrometric data of maximum discharges shows that in more than 80% of all cases, such discharges occur during the spring-summer period (Table 1). Compared to the southern slope rivers, the rivers of the Northern Caucasus exhibit high flows in the spring and such flows are formed by the rainwater, but seldom exceed the maxima of the spring-summer period.

Annually in spring and summer, as a result of increased temperatures, torrential rains are observed on numerous mountain rivers and contribute to powerful floods. In the country there is a general system of implementation of flood protection measures, but more than one hundred villages and towns and their infrastructures are exposed to harmful impacts of floods and high waters of the mountain rivers.

Large scale deforestation all over the South Caucasus is the driving force behind the erosion, sediment export and sedimentation problems in the affected countries. As a result, dam reservoirs are increasingly filling up with sediment, which contributes to low water quality and flooding upstream and downstream (Khalilov 2003).

In general, the irrigation reservoirs are not primarily designed to act as flood storage reservoirs. However, most of the existing reservoirs are able to partially mitigate the flood effects, which otherwise would be devastating. These reservoirs act as flood storage reservoirs whenever they have ample (dynamic) storage capacity compared to the volume of the spring runoff of that particular river. In other rivers (or river sections) no reservoirs exist, e.g. on the Aras River, where along both riverbanks high dykes were constructed to prevent inundation of homes, farms and arable land. Since many spillway structures of dam reservoirs have rather low capacities, the dynamic flood storage capacity is actually larger than it should be under proper spillway design conditions.

Apart from inundation of infrastructure and cultivated fields, the erosive forces of river floods tend to damage riverbanks. These effects may be mitigated by the construction of groins or other bank protection measures, which have been designed and constructed along various rivers.

In establishing the design volumes of storage reservoirs and rules for their operation, the designers and operators should also take into account the possibility of flood storage. It is proposed that in establishing operation rules of the irrigation reservoirs, flood water retention should be considered and verified in order to generate optimum multiple benefits.

At the same magnitude as deforestation, flood irrigation is inducing high sediment loads, causing sedimentation downstream and therefore flooding in the river delta. Also the irrigation canals themselves are susceptible to blockage by sediment and consequently require more intensive maintenance (Verdiyev 2002). The sediment load also causes problems in drinking water supply.

Table 1. Average values of the characteristics of a spring flood (Rustamov and Gashgay 1989; Mamedov 1989; Mamedov 2002; Verdiyev 2002)

№	River - location	Water catchment area		Spring flood				Average annual runoff, m^3/s
		Area, km^2	Average elevation, m	Dates (day, month)		Duration, days	Depth of runoff, mm	
				The beginning	The end			
1	Kura-Gragkesaman	35,900		(12.03)	(20.06)	(99)	(110)	210
2	Kura-Zardob	76,000		(15.03)	(20.06)	(96)	(82)	334
3	Kura-Salyan	188,000		(17.03)	(21.06)	(94)	(49)	479
4	Ganikh(Alazani)-Agrichay	11,600		(10.03)	(28.06)	(111)	(246)	110
5	Gabirri (Iori)-Kesaman	4,270		(7.03)	(29.06)	(113)	(17)	3.94
6	Aras-Saatli	100,000		19.03	28.06	101	40	138
7	Agrichay-near a mouth	1,810	1040	(16.03)	(20.06)	(95)	(186)	17.8
8	Turyanchay-hydrounit	1,440	1180	(9.03)	(24.06)	(106)	(93)	7.23
9	Gekchay-Gekchay	1,480	970	8.03	21.06	114	108	12.8

10	Girdimanchay-Karanour	353	1820	10.03	30.06	113	319	6.83
11	Akstafachay-Barkhudarli	1,763		(14.03)	(29.06)	(106)	(92)	8.65
12	Shamkirchay-Barsum	922	1900	(7.03)	(2.08)	(153)	(285)	13.9
13	Ganjachay-Zurnabad	314	2090	08.03	07.08	152	282	4.25
14	Kurakchay-Dozular	439	1770	12.03	15.08	156	176	3.97
15	Terter-Madagiz	2,460	2030	13.03	10.08	151	190	18.1
16	Nakhchivanchay-Karababa	449	2060	13.03	25.06	106	268	4.83
17	Gilanchay-Bilav	299	2360	19.03	13.07	116	284	4.19
18	Akera-Lachin (Abdalyar)	1,180	2130	27.03	10.07	106	148	10.5

Table 2. Maximum streamflow of the rivers (Rustamov and Gashgay 1989; Mamedov 1989; Mamedov 2002; Verdiyev 2002)

№	River - location	The period of observation, years	Reservoir		Record peak discharge		Average water discharge and specific flow	
			Area, km^2	Average elevation, m	m^3/s	Day, month, year	m^3/s	l/s.km^2
1	Kura-Gragkesaman	1953-58 1986-95	3,590		2,190	12.06.87	283	7.88
2	Kura-Zardob	1953-95	76,000		1,240	02-04.07.78	334	4.39
3	Kura-Salyan	1953-95	188,000		2,350	11.05.69	477	2.53
4	Ganikh (Alazani)-Agrichay	1950-95	11,600		684	30.04.82	110	9.48
5	Gabirri(Iori)-Kesaman	1976-90, 1992-94	4,270		92.8	11.04.84	3.94	0.92
6	Aras-Saatli	1971-95	100,000		1,480	26.05.76	138	1.38
7	Agrichay-near a mouth	1962-95	1,810	1040	141	06.05.67	17.8	9.83

8	Turyanchay-hydrounit	1968-81, 1984-95	1,420	1120	170	26.06.77	7.23	5.02
9	Gekchay-Gekchay	1929, 1930, 1934-42, 1947-93, 1995	1,480	970	424	07.07.63	12.8	8.64
10	Girdimanchay-Karanour	1966-88, 1990-92, 1994	352	1820	201	15.07.88	6.83	19.3
11	Akstafachay-Barkhudarli	1974-91	1,763		85.5	13.04.87	8.65	4.91
12	Shamkirchay-Barsum	1928-88	922	1900	130	07.07.83	13.9	15.1
13	Ganjachay-Zurnabad	1928-95	314	2090	107	11.07.65	4.25	13.5
14	Kurakchay-Dozular	1939-46, 1948-95	439	1770	168	02.07.58	3.97	9.04
15	Terter-Madagiz	1925-73, 1976, 1977	2,460	2030	647	30.06.40	22.4	9.1
16	Nakhchivanchay-Karababa	1940-43, 1948-95	449	2060	211	02.07.60	4.83	10.7
17	Gilanchay-Bilav	1961-95	299	2360	107	28.04.69	4.19	14.0
18	Akera-Lachin (Abdalyar)	1927, 1928, 1932-35, 1952-87	1,180	2130	106	14.05.74	10.5	8.8

References

Rustamov SG, Gashgay RM (1989) Water Resources of Azerbaijan SSR, Baku

Mamedov MA (1989) Accounts of the maximum discharges of the mountain rivers, Leningrad

Khalilov SB (2003) Reservoirs of Azerbaijan and their ecological problems. Baku

Tacis Project (2003) Tacis joint river basin management project (Kura Basin): progress report

Fatullayev GY (2002) Modern change of water resources and water mode of the rivers of Southern Caucasus. Baku

Mamedov MA (2002) Hydrography of Azerbaijan. Baku

Verdiyev RG (2002) Water resources of the rivers of East Caucasus under conditions of climate change. Baku

Imanov FA (2000) The minimum flows of the rivers of Caucasus. Baku

Museyibov MA (1998) Physical geography of Azerbaijan. Baku

FLOOD DEFENCE BY MEANS OF COMPLEX STRUCTURAL MEASURES

STEFANO PAGLIARA[1]
University of Pisa, Pisa, Italy 56126

Abstract. This paper analyses the complexity of using different structural measures in the flood risk mitigation planning process for river basins divided into different administrative jurisdictions. In particular the case of complex systems of detention basins is studied. Detention basins are widely used in Italy. Problems arise between local and regional communities when setting the operation of such structures. Another conflict arises due to the fact that basins built on tributaries are often designed in a way that creates local benefits only; sometimes, such projects increase the flood risk downstream. Also, a series of detention or retarding basins must be planned well in order to obtain the maximum benefit during the period of implementation of the master plan that can last for some decades.

Keywords: flood defence, mitigation measures, flood risk, detention basins.

1. Introduction

Flood defence may be a complex task if the mitigation measures are spread out in a large basin and if different communities are involved in the decision making process.

Only the approach encompassing the study of the whole basin guarantees the effectiveness of structural measures; that means that a specific measure produces benefits both for the local community as well as the downstream communities.

---

[1] To whom correspondence should be addressed. Stefano Pagliara, University of Pisa, Department of Civil Engineering, v. Gabba, 22 56126 Pisa, Italy; e-mail: s.pagliara@ing.unipi.it

In Italy, the legislation (Law 183/89) stipulates the presence of a unique "River Authority" in each river basin. This means that many problems regarding transboundary floods (in the sense that many regional and local administrations are involved) are easier to solve. However, many conflicts may arise due to the fact that the local communities try to obtain the maximum local benefit from the implementation of a single structural flood protection measure, which may often generate bigger flood peaks downstream. Again, the interaction among many structural protection measures can be very complex and cause an increase of the flood peak.

The paper presents some examples of a river flood protection master plan. The choice of the Tuscany region authorities has been to build detention basins rather than dam reservoirs in order to contain the design flood; this is mainly due to environmental and historical reasons. Another design concept consists in setting the design flood return period at 200 years which represents a very high level of protection. The last widely accepted concept states that no flood protection work is allowed to increase the risk of downstream flooding.

2. Methods

Two different basins were chosen to show that the complexity of the flooding problem is, to some extent, independent of the basin size; the first case study is a small catchment (the Cornia river) located in the south of Tuscany, while the second one is a large basin (the Arno river) located in the northern part of the region and draining into the Thyrrenian sea.

2.1. CORNIA RIVER BASIN

The Cornia river flows into the Thyrrenian sea and has a catchment area of about 350 km^2; this river flooded large areas in the past and has a system of flood protection dykes. The flow capacity of the downstream reach can not contain discharges with return periods greater than 60-70 years (i.e., > 900 m^3/s).

Figure 1 shows a layout of the whole catchment. The most downstream part of the river is an isolated reach because of the presence of a dyke system and flat relief of the area. The upper part of the basin has a mean slope of about 20%.

The downstream part of the river is susceptible to high hydraulic risk. Figure 2 shows the 200-year hydrograph in the last 8 km where the river is confined by dykes and where urban areas are located; it is evident that flood discharge is reduced in this reach by about 400 m^3/s due to the overtopping of dykes and overflow of about 6-8 millions of cubic metres of flood water into the adjacent urban areas. To deal with this problem, various structural measures have been planned, with the final choice being building of detention basins and the strengthening of the dyke system.

Fig. 1. Cornia river catchment (note the isolated nature of the river downstream reach caused by the presence of a dyke system)

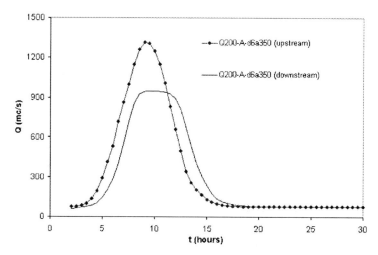

Fig. 2. Cornia river: hydrographs for a return period of 200 years

The first problem caused by the "transboundary" issues is due to the fact that the urbanised areas are all located downstream, but the detention basins are upstream. This results in a big controversy between local governments in terms of their interests.

Figure 3 shows the structural measures necessary for the mitigation of the hydraulic (flood) risk with a total cost of about 27 million of €, or about 2.7 €/m^3 of stored water. The system of sub-catchments is rather complex and various comparable master plan scenarios have been proposed in order to solve the flooding problems. Each scenario considers a different distribution of the sub-catchments. This is caused by the fact that each local administration will have not more than 1 or 2 sub-catchments on its territory. In fact, flood retention basins use up space in flood plains, which despite of their flood-prone nature represent the zones where all the administrations would like to plan industrial or commercial or even residential developments.

Figure 4 shows the system of earthen dykes, which need to be strengthened to prevent their breach due to overtopping. The reason for this measure is that even though there is an urgency to defend the main urban areas, it is not possible to elevate the dykes because of the increased risk downstream.

Finally, consideration of various flood scenarios is necessary in order to compare the actual situation and the effect of the different mitigation measures to be completed in different years (i.e., the structural measures will be built over a certain period of time).

Figure 5 shows the points at which overtopping and/or dyke breach occur for the 200-year flood event. Figure 6 shows, for the downstream part of the basin, the water depth contours for the 200-year flood event. A total of about 6-8 millions m^3 will overflow onto the flood plain.

Figure 7 shows the difference in the flooded areas due to the strengthening of the dykes; protection of the dykes against breaching during the overtopping process dramatically diminishes the volume of water inundating the area.

Figure 8 displays flow hydrographs in the downstream part of the river for two different planning scenarios and different rainfall durations with a 200-year return period. It is evident that both scenarios can reduce the flood peak from 1,300 to 800 m^3/s, with the lower flow representing the discharge that can be conveyed by the river with the required freeboard.

Finally, it is important to establish the correct priority in building the structural measures. For the Cornia river catchment the maximum benefit has been found by building two flood retention reservoirs, which are shown in Fig. 9 that. In this scheme, flooding is eliminated even without the freeboard in the downstream reach.

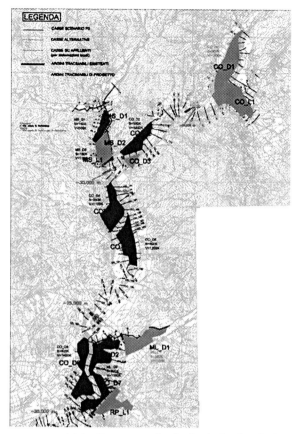

Fig. 3. Grey zones represent detention or retention basins in the Cornia river catchment

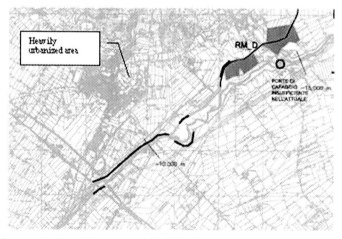

Fig. 4. Reinforcement of the dyke system (black lines)

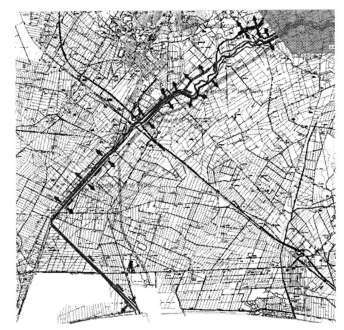

Fig. 5. Flooding points (arrows show the points of dyke overtopping and/or breach)

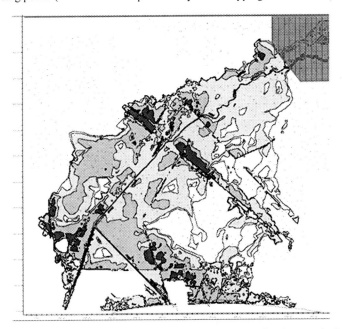

Fig. 6. Contours of the maximum water depths during the 200-year flood (grey shades represent various water depths)

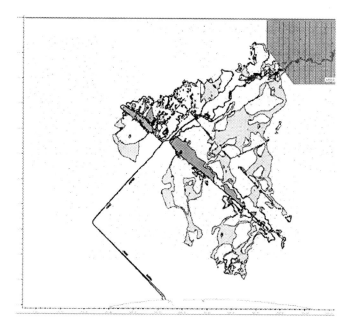

Fig. 7. Difference between the 200-year flooding with and without dyke reinforcement (grey shades show different water depths)

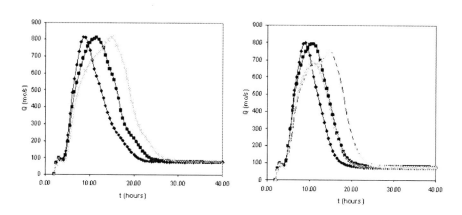

Fig. 8. Flow hydrographs in the most downstream reach of the river for two scenarios, different rainfall durations and a 200-year return period

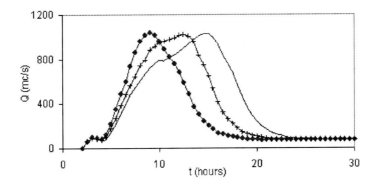

Fig. 9. Flow hydrographs in the most downstream reach for different rainfall durations, a return period of 200 years, and with two flood storage reservoirs built

2.2. ARNO RIVER BASIN

The Arno River has a basin of more than 8,000 km^2 and represents a large river in the Tuscany region with a long history of flooding; the last big flood occurred in 1966, when the city of Florence was flooded and large damages occurred in the whole river valley (Pagliara 2004). Figure 10 shows the beginning of the flood event in Florence and Figure 11 shows the river basin.

The "master plan for the reduction of the hydraulic risk" expects implementation of structural measures costing many thousands of millions of Euro. This means that the plan will be realised over many decades. A question arises: if we know the current flood risk in the basin and the risk level after the master plan implementation, what will be the risk in the meantime?

This question is complex, because during the plan implementation, the overall risk will generally diminish, but locally it can increase. And again, which are the structural measures that maximise the benefits and which should receive priority? To answer these questions, the river Authority has built a mathematical model in which many different scenarios can be tested. The model results are really complex and their implementation needs a very delicate planning process in order to satisfy the many and often conflicting interests of local administrations operating at different levels. The master plan foresees retention basins for about 300 millions m^3, located partly along the main river and partly on the major tributaries, dyke improvements and other measures. When implemented, the flood protection system should prevent damages for floods with return periods estimated between 150 and 200 years.

Fig. 10. Image of the big flood on the Arno river in Florence (Italy) in 1966

Fig. 11. Map of the Arno river basin

One result that is not self-evident is that during the plan implementation, the storage reservoirs built on the main river often tend to increase flood peaks downstream. This is due to the fact that such reservoirs store only part of the flood volume that would overtop the dyke system.

Also, the 200-year peak discharge for a reach located in the middle of the basin is about 4,000 m³/s, while at the lower end of the basin it is 2,700 m³/s. This means that the theoretical 200-year flood for the whole basin could be about 7,000 m³/s and also that a large part of the flood volume is generated in the middle part of the basin. Consequently, storage introduced by the planned reservoir system just compensates for the water normally spilling over dykes. This is evident in Figure 12 where the hydrograph in the downstream part of the Arno river is shown for different conditions (the actual one and those corresponding to different scenarios of planned measures). The flood peak practically remains the same despite the fact that many storage reservoirs are built upstream.

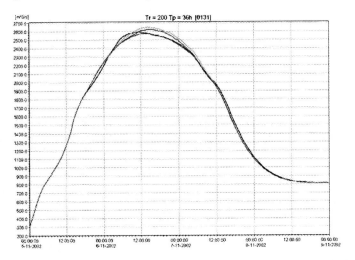

Fig. 12. Hydrographs at the downstream end of the Arno river for a return period of 200 years and different scenarios

3. Conclusions

Two examples of complex river basin master plans for the reduction of the flood (hydraulic) risk have been analysed. For both the Cornia and the Arno river basins a great number of structural flood protection measures need careful planning with respect to spatial and temporal distributions, and the involvement of all stakeholders, including the local administrators, to obtain the expected outcomes.

References

Pagliara S (2001) Hydraulic modelling and historical inundation assessment for the Versilia River. In: Glade T et al. (eds) The use of historical data in natural hazard assessment, Kluwer Academic Publishers, Netherlands, pp 141-150

Pagliara S (2004) Flood risk mitigation and management. Proc. Of the NATO ARW on "Flood risk management, hazards, vulnerability, mitigation measures" Ostrov, Czech Rep., Oct. 6-10, pp149-158

Rossi, Harmancioglu, Yevjevich (eds) (1992) Coping with floods. NATO ASI Proc. Advanced Study Institute in Erice, Nov. 3-15

Yen BC (1995) Urban flood hazard and its mitigation. In: Cheng FY, Sheu MS (eds) Urban disaster mitigation, Elsevier Science Ltd.

DYKE FAILURES IN HUNGARY OF THE PAST 220 YEARS

SÁNDOR TÓTH[1]
National Directorate for Environment, Nature Conservation and Water, Budapest, Hungary

LÁSZLÓ NAGY
Budapest University of Technology, Department of Geotechnics, Hungary

Abstract. Historical research covering the past 220 years has identified 1,245 dyke breaches in Hungary, including 556 cases where the length of the breach was known. During the past 50 years (1954-2004), the dominant type of failure has shifted from overtopping to hydraulic subsoil failure, due to the continuous heightening of the dykes.

Keywords: flood hazard, dyke breach, dyke failure formation process, location factor, overtopping, hydraulic failure.

1. Introduction

1.1. FLOOD HAZARD IN HUNGARY

Under the particular physiographic conditions of the Carpathian basin, important and steadily growing interests have been paid to flood defence for centuries. The fundamental cause of the grave flood hazard is that the overwhelmingly plain country is situated in the deepest part of the Carpathian basin, where the flood waves rushing down from the surrounding Carpathian and Alpine headwater catchments are slowed down, combine and coincide with each other resulting often in high river stages of extended duration. Owing to the climate and the physiographic

---

[1] To whom all correspondence should be addressed. Sandor Toth, National Directorate for Environment, Nature Conservation and Water, Budapest, Hungary; e-mail: toth.sandor@ovf.hu

situation floods are likely to occur practically on any Hungarian river in any season of the year. The largest tributary to the Danube is the Tisza River draining the eastern part of the Carpathian basin.

On the Hungarian territory, only the Rába and Rábca, Mura and Dráva (forming national border with Slovenia and Croatia), the tributaries to the Danube, and Szamos, Bodrog, Hármas-Körös (Triple-Körös, integrating Sebes-, Kettős /Double/-Körös as a receiving stream of Fekete /Black/- and Fehér /White/-Körös) and Maros in the Tisza Valley can be considered as larger rivers. Characteristic discharges of these rivers are summarised in Table 1. Smaller rivers include the Sajó, Hernád, Tarna and Zagyva in the Tisza Valley, and the Sió, another tributary of the Danube.

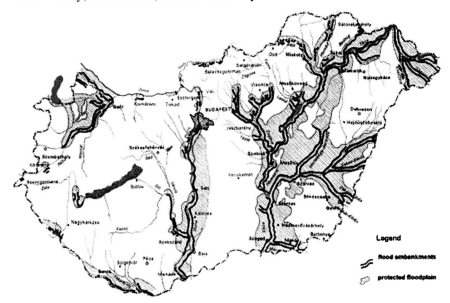

Fig. 1. Main rivers, floodplains and dykes in Hungary

Table 1. Characteristics of Hungarian rivers

River	Gauging station	Q_{min}	Q_{max}	Q_{max}/Q_{min}
		m^3/s		
Duna	Rajka	570.0	10,300	18
Duna	Budapest	580.0	8,600	14
Duna	Mohács	618.0	7,850	13
Rába	Sárvár	6.5	800	123
Ipoly	Balassagyarmat	3.5	360	103
Dráva	Barcs	185.0	3,050	17

River	Gauging station	Q_{min}	Q_{max}	Q_{max}/Q_{min}
Tisza	Záhony	47.0	3,750	80
Tisza	Szolnok	60.0	3,820	64
Tisza	Szeged	95.0	4,700	49
Szamos	Csenger	8.0	1,350	169
Bodrog	Sárospatak	2.0	1,250	313
Sajó	Felsőzsolca	2.4	545	227
Fehér-Körös	Gyula	1.0	610	610
Fekete-Körös	Sarkad	1.0	810	810
Kettős-Körös	Békés	2.3	905	393
Hármas-Körös	Gyoma	4.5	1,800	400
Maros	Makó	22.0	2,450	111

1.2. INVESTIGATION OF DYKE FAILURES

In the framework of the IMPACT integrated project under FP5 of the European Commission in 2002-2004, Hungarian investigators from HEURAqua Ltd., as project partners, carried out an extensive investigation of dyke failures during the past 220 years in Hungary. The number of dike failures identified in Hungary by historical studies is 1,245, including 559 (45%) with a known length of breach. Table 2 shows the breakdown of the number and length of breaches by individual rivers. Different authors recorded alternative lengths for 6 of the 559 dike breaches of identified length.

Table 2. Number and length of breaches

River	Dyke breach	Identified length		Total length	Average length
	Number	Number	%	(m)	(m)
Danube	270	78	29	8,189	104
Tisza	219	96	44	9,192	95
Tributaries	539	290	54	12,422	42
Small rivers	217	95	44	3,424	35
Total:	1,245	559	45	33,227	59

Distribution of the failure mechanisms of the breaches in those cases when the causes of the dyke breach could be identified is shown in Fig. 2.

The graph clearly shows that the majority of the breaches were caused by overtopping, which led to the continuous raising and reinforcement of the dykes, as can be seen in Fig. 3.

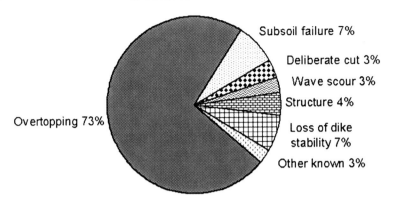

Fig. 2. Distribution of identified failure mechanisms of the breaches of Hungarian flood dykes (1802-2004)

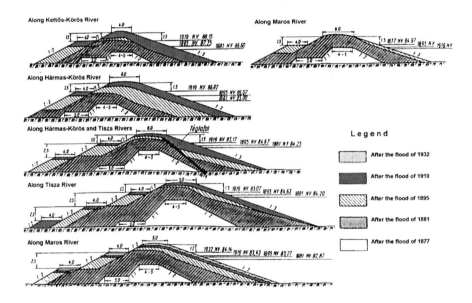

Fig. 3. Dykes with standard cross-sections

2. Dyke failures of the past 50 years in Hungary

The floods during the past 50 years (1954-2004) have caused 117 embankment failures, of which 72 were caused by overtopping, 21 by hydraulic soil failure, 10 by stability loss of the embankment, and 2 by leakage along structures; the failure mechanism could not be identified in 12 cases (Fig. 4). Although the number or the proportion of overtopping is still quite high, it can be explained by the 1956 ice-jam flood on the Danube

(33 cases), the flood in 1963 along the relatively small Tarna river and its tributaries originating in the Mátra mountains (16), and the extreme floods on smaller tributaries of the Danube, the Répce and Lajta, in 1965 causing 8 cases (while the Great Flood of the Danube was successfully defended in Hungary). Further 4 cases occurred in 1972 along the river Mura, 2 in 1974 along the torrential Fekete-Körös, another one on the same river in 1981, and finally the extreme flood of the Upper-Tisza in 2001, which resulted in 2 breaches along the Tisza right bank in Bereg and another 2 along its tributary, the Túr.

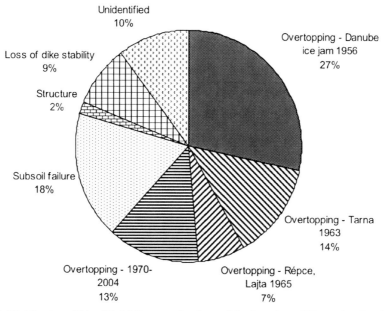

Fig. 4. Distribution of identified failure mechanisms of the breaches of Hungarian flood dykes (1954-2004)

As a consequence of the continuous heightening of the dykes, the type of failure has changed. While in the earlier historical period of flood protection the main danger and cause of damages was overtopping of the river dykes, which was followed by the complete erosion of the dyke body itself, the heightening of the dykes led to growing exposure of the foundation soil contributing to the development of hydraulic soil failure, boils and piping, while the growing duration of floods raised the risk of dyke saturation (Fig. 4).

The operation of reservoir systems (cascades) has also modified the falling limb of flood hydrographs (the water levels are falling at faster rates), which may cause slumping of the waterside banks of the dykes and natural rivers.

3. Variation of the length of dyke breaches in Hungary

3.1. VARIATION OF THE LENGTH OF DYKE FAILURES ALONG THE RIVER

The first aspect of studying the length of dyke failures is to see whether or not the longitudinal profile of the affected river shows some alteration or regularity. The Tisza River, which is 945 km long, flows between dykes along an 800 km stretch downstream from Huszt (Khust, Ukraine). In Hungary the Tisza River flows over a length of approximately 600 km between chainage 153 and 744 km sections. Fig. 5 presents the length of dyke failures distributed by the profiling of the Tisza River. The average length of breaches along the Upper-Tisza does not deviate significantly from those of the Lower-Tisza, although both the linear and exponential trend lines show a slight rise towards the receiving river, e.g. the Danube, but such an increase is not significant. The exponential trend line indicates the effect of the Danube, however, it should be mentioned that the backwater effect of the receiving river is felt along a longer reach.

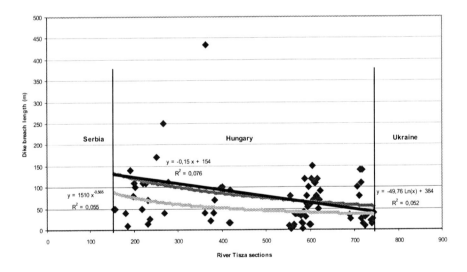

Fig. 5. Dyke failure length along the longitudinal section of the Tisza

A logarithmic function trend reveals that the width of dyke breaches occurred along the Danube does not vary significantly with the location of the breach, whilst an exponential trend shows that smaller widths are found in the upstream areas (Fig. 6). Plots by longitudinal sections are possible only for the Danube and the Tisza rivers.

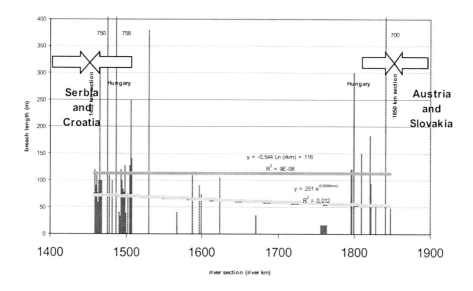

Fig. 6. Dyke failure length along the longitudinal section of the Danube

3.2. EFFECT OF THE OVERFLOW HEIGHT ON THE LENGTH OF DYKE BREACHES

We estimated the head over the overflow crest at the initial stage of the failure for dyke breaches along the Tisza. These estimates should be treated with caution because the accuracy of the overflow height may be within ± 0.3 m of the actual value, due to, among other things, the aforementioned changes of the river and the fact that the high water gradient of the Tisza is less than 3 cm/km in some locations and more than 1 m/km in other locations.

Figure 7 shows nothing more than just a tendency of the relationship between the head of overtopping and the breach length. As the overflow height increases, so does the length of the dyke breach but the correlation is poor in terms of both the power and exponential functions (Fig. 7). That is probably due to the multitude factors that are here at play. All in all, the results do not contradict the physical law that raising the height of overflow will increase the boundary shear of the water which corresponds to the increase of the opening of a dyke failure. The data show that the lower the height of the dyke, the less variable the width of dyke breaches will be. The two points in the left hand side of Fig. 7 indicate high ground overflows.

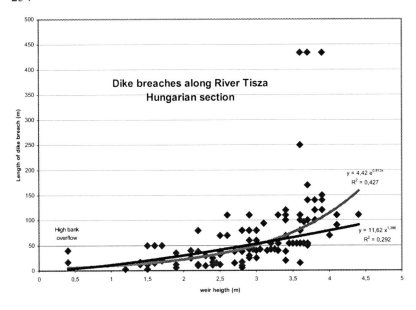

Fig. 7a. Dyke breach length as a function of the overflow height along the Tisza River

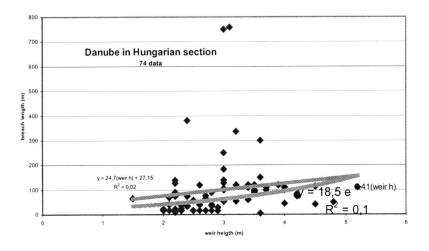

Fig. 7b. Dyke failure length as a function of the overflow height along the Danube River

3.3. EFFECT OF THE GEOTECHNICAL PROPERTIES OF THE DYKES

Results of the analysis of the influence of soil types are summarised in Table 3 and Figure 8. The hazard of shifting particles, erosion and hence of an increasing length of a dyke breach is higher with fine-particle non-cohesive soils than in the case of clays and coarse grain gravel (note, that the latter soils are rarely used for in construction of homogeneous dykes).

Table 3. Final length of breach for different soil types and dyke heights

Soil type	Data N	Breach length (m)			Dyke height (m)		
		min	average	max	min	average	max
CH	4	66	96	145	3.0	5.5	11.1
SC	17	5	87	120	3.5	7.6	14.9
CS	10	22	107	250	2.4	7.0	18.2
MS	4	58	135	210	2.5	4.0	8.0
S, GS, MG	4	35	92	184	6.6	8.2	11.3
Total	39						

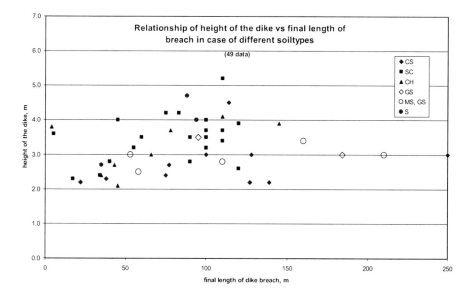

Fig. 8. Relationship of dyke height vs the final breach length for different soil types

Fine particulate soils with no (or low) cohesion can be dislocated and washed away at lower speeds of water than the highly cohesive soils. This effect is important at the edges of breach openings.

Practical experience shows that a vertical wall of earth is normally formed at the edge of a dyke breach by the dyke stub. The water flowing by wears away the dyke toe and the dyke above will collapse and leave behind another vertical wall. The faster the flowing water can destruct the edge of the dyke stub the faster this process proceeds. If a dyke stub is made of a soil that even slower water flows can also destruct, the width of a dyke breach will be larger. Such soils include the types that are easily erodible, poorly compacted soils and dispersed soils.

3.4. EFFECT OF THE RIVER DISCHARGE ON THE LENGTH OF DYKE BREACHES

The average width of the breach of the 125 dyke failures with the known breach length from the total number of 397 dyke failures along the Danube Hungarian section was 130 metres (Figure 9). The variation of the data taken from the left and right hand sides differs, which is why the areas delineated at the top and the bottom of the output square in Figure 9 show an identical variance about the average.

Figures 10 and 11 show the average breach width for the four categories mentioned above in linear and semi-logarithmic scales. It is clearly visible that, on average, rivers with larger discharges form larger dyke breaches. If one can draw a conclusion from the four figures (Figs. 8-11), one might say that the width of a dyke breach increases with an increasing river discharge.

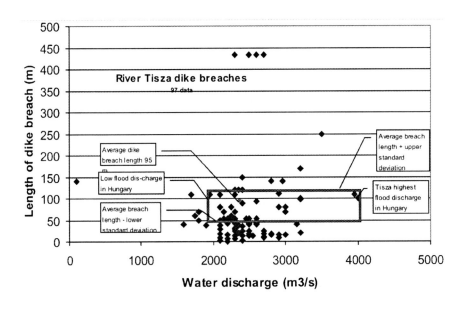

Fig. 9. Dyke failure length as a function of discharge along the Tisza River

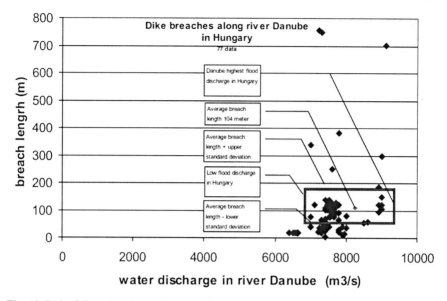

Fig. 10. Dyke failure length as a function of discharge along the Danube River

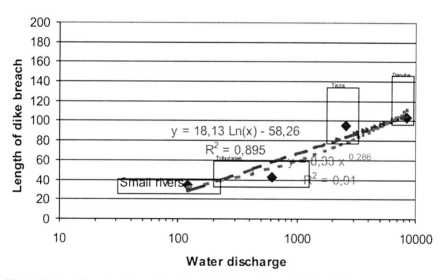

Fig. 11. Dyke failure length as a function of discharge (logarithmic scale)

4. Summary

Flood fighting activities in preceding centuries and systematic research have produced a collection of 1,245 historical data regarding dyke failures in Hungary. Despite the gaps and frequent errors in historical data, the high number of dyke breaches facilitates statistical processing of data and the evaluation of the results allows us to draw interesting conclusions and lessons for future generations, for instance regarding the length of dyke breaches. The effect of human intervention is easy to trace in the system of flood control on the basis of the changing number and length of dyke breaches.

The expected length of a dyke breach depends on a number of interrelated factors, yet there used to be no method for estimation of such values. The present study allows us to state that a starting point has been created for increasing the accuracy of estimating expected breach lengths. More than 1,000 of historical data points have been processed to calculate the average breach length in the dykes of the Danube, the Tisza, their tributaries and some smaller Hungarian rivers. Neither the breach length results, nor the temporal trend of breaches contradict the laws of physics.

Although quite a few factors, such as the geotechnical properties of dykes, the flow conditions of rivers relative to the location of the dyke failure, and the effect of protected side terrain conditions, are difficult to express quantitatively, for other factors, such as the flood fighting activities, quantitative expressions were impossible. Consequently, statistical processing offered results which were easy to interpret regarding a river or a river type. The analyses performed allow us to make technically sound estimates of expected breach lengths along certain rivers or river sections, and these estimates can be used in site-specific calculations.

4.1. RESULTS FOR THE INDUSTRY

We do believe that the above results give very important data for the industry with respect to their preparedness and emergency management activities.

Relationships between the lengths of breaches and the height of overflow, and the river discharge and the location of the breach along the river, may offer substantial support in planning and implementing confinement activities.

References

Nagy L, Toth S (2004) IMPACT (Investigation of Extreme Flood Processes & Uncertainty) Project (Contract No.: EVG1-CT-2001-00037) WP6 Monitoring and Case Study, Detailed Technical Report on the collation and analysis of dike breach data with regards to formation process and location factors (Conclusions). National Directorate for Environment, Nature Conservation and Water, Budapest, Hungary

EFFECT OF THE SALARD TEMPORARY STORAGE RESERVOIR ON THE BARCAU RIVER FLOWS NEAR THE ROMANIA-HUNGARY BORDER

ADRIAN PURDEL[1]
National Institute of Hydrology and Water Management, Bucharest, Romania 031686

PETRE MAZILU
S. C. AQUAPROIECT S.A., Bucharest, Romania

Abstract. The Barcau River is a transboundary river, which forms a component of the Cris hydrographic basin. The river is nearly 134 km long and drains an area of 1,960 km^2 on the territory of Romania. The shape of the hydrographic basin, the presence of the soils with reduced permeability, intensive rainfalls and the low conveyance capacity of the minor Barcau riverbed as well as of its tributaries, are the major causes of frequent floods. The largest floods were recorded in 1970, 1974, 1980 and 1997. The existing flood protection works throughout the hydrographic basin consist mostly of dykes (about 101 km long, on either riverbank) and bed regulation works (nearly 134 km long). They correspond to the flood protection storage volumes provided in local storage reservoirs situated on the main tributaries, and a temporary storage reservoir, of the polder type, situated right upstream the Salard municipality. First, the main river course was dyked - along a 17 km distance between the Romanian- Hungarian border and Salard (dykes continue on the Hungarian territory up to the Tisa River) in order to protect the settlements and agricultural lands in the lower basin against floods. The construction of dykes was continued between 1985 and 1989 on both banks of the Barcau River, in the upstream direction, to the town of Marghita. In order to compensate for flood build

---

[1] To whom correspondence should be addressed. Adrian Purdel, National Institute of Hydrology and water Management, 97 Sos. Bucuresti-Ploiesti, sector 1, 013686, Bucharest, Romania; e-mail: eu_andy@yahoo.com

up due to these works, the Salard temporary storage reservoir was built. The hydraulic calculations concerning the flood relief provided by the Salard polder in the reach of interest to both Romania and Hungary, between the Salard and Berettyoujfalu hydrometric stations, were discussed at the meetings held by the Romanian-Hungarian Sub-commission on flood protection in 2004. The hydraulic calculations were conducted by the Romanian and Hungarian teams, using the numerical HEC-RAS model applied to the same sets of hydraulic and topographic data. Following calibration of the mathematical model on the greatest floods recorded, it was proved that the bed roughness coefficients increased in time. This led to cleaning the riverbed. Having calibrated the mathematical model, the influence of the Salard polder on flow in the river reach of common concern was demonstrated in various scenarios, which would be implemented in the upstream area of the Barcau.

Keywords: Transboundary River, polder, dams, HEC – RAS model

1. Introduction

The Barcau River Basin has an area of 1,960 km^2 in Romania. The river length is about 134 km, and the mean slope is about 0.4%; in the upper river reach, the slope is about 0.7%, and 0.057% in the downstream border reach.

To protect inhabited and agricultural areas in the Barcau River Basin, first 17 km of dykes were built. Dyke construction was continued on the Hungarian territory up to the Tisa River. The dykes are 3-4 m high and have a distance between them of about 100 m. These design parameters insure conveyance of the maximum discharge of 400 m^3/s.

Between 1985 and 1990 the work continued in the upstream direction in the sector Salard – Balc mainly by building dykes on one or both banks. The dyke heights were established to provide a 35 cm freeboard for flows with flood probability > 5%.

For protection of the town of Marghita and its surroundings, a weir was built so as to afford an attenuation of floods generated by the above-mentioned construction.

Due to the hydraulic structures that have been built for flood defence in the Marghita – Salard reach, when the incoming flow corresponds to a 5% flooding probability, the maximum discharge downstream of Salard would reach, due to the flow build up, a value corresponding to the 1% flooding probability. In order to compensate (partly) for this effect and to address the border conditions, the Salard temporary storage reservoir was built. It will ensure the conveyance of a flow corresponding to the maximum transport capacity of the riverbed.

The Salard temporary storage reservoir has been built according to the Romanian regulations and is located on the left bank of the Barcau River. The inlet weir is 122.5 m wide and the elevation of the weir crest is 110.86 m ASL.

According to the hydraulic calculations presented by AQUAPROIECT (Mazilu and Mustatea 2002), water inflow into the temporary storage reservoir is allowed when the flood discharge exceeds the value corresponding to the 10% discharge probability ($Q_{10\%}$) under the conditions of the flow regime modified by dykes.

2. Data and mathematical models used in analysis

The hydraulic calculations for the Barcau River, intended to show the influence of the Salard temporary storage reservoir in the analysed reach of common interest to Romania and Hungary (between the Salard and Berettyoujfalu gauging stations) were done with the U.S. numerical model HEC-RAS and with the Romanian models MIPE and UNDA. These models simulate flash floods in natural and modified flow regime conditions. The length of the analysed reach was 48.7 km, with 17.7 km in Romania and 31 km in Hungary.

For calibrating these models the following input data were used:

- hydrological data (recorded flow and stage, Streng et al. 2001); river bed field data; and, hydrographs with various probabilities of occurrence,
- topographical data: cross-sections, longitudinal-sections, topographical plans, data regarding reference systems (provided by Mazilu and Mustatea, (2002) and by the Hungarian team), and
- data for hydraulic structures.

For hydraulic calculations of unsteady flow, the calibration of the proposed mathematical models is achieved by adjusting riverbed parameters and coefficients. The boundary conditions are imposed by the rating curve at the downstream reach end and the input hydrograph in the upstream reach end.

The parameters that can be modified during the calibration process are: Manning n coefficients (which model bed roughness), Figure 1; the length of the major riverbed storage areas along the main flood axis; the determination of the zones where the major riverbed flood begins, local depression areas; finding and modelling reverse flow areas; accurate model coefficients calibrated by optimising the time and the length of computational steps, etc. On the Hungarian territory the analysed river reach was divided into two sub-reaches with different n coefficients.

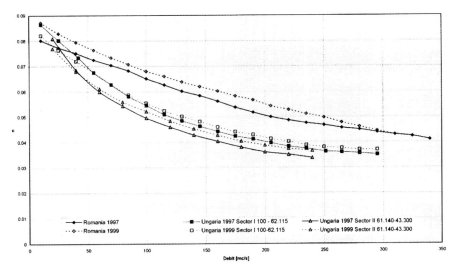

Fig. 1. Variation of Manning coefficients with discharge

3. The influence of the Salard temporary storage reservoir on the flood regime in the Salard – Berettyyoujfalu reach

In the border section, according to the agreements between the Romanian and Hungarian parties, the imposed limiting condition is not to exceed the maximum transport capacity of the downstream channel, which is about 400 m³/s. Such a value is close to the 1% frequency discharge (370 m³/s) at the Salard gauging station.

The results of hydraulic calculations regarding the influence of the Salard temporary storage reservoir on the Salard – Berettyoujfalu reach, for the flood with a 44.2 mm runoff depth corresponding to the 1% discharge probability and different scenarios, are presented below.

- *Scenario 1*: The Barcau river bed with the existing dykes (border – Salard - 17 km and Salard - Almasu Mic - 45 km) without the flood attenuation works, and the Salard temporary storage reservoir, the Marghita temporary storage reservoir and the Suplacu de Barcau reservoir. Calculations took into account that the flow passes between the dykes (class II standard – 1% exceeding probability) in the Almasu Mic border reach. The maximum discharge for 1% probability at the Salard gauging station is in this case 450 m³/s, compared to 370 m³/s in the natural regime. At the border section the maximum discharge would be 446 m³/s and at Berettyyoujfalu around 429 m³/s. After the passage of this flood through the Salard – Berettyoujfalu reach, it was determined that the existing dykes do not have sufficient conveyance capacity, and the flood exceeds the left bank dykes on the Hungarian territory.

- *Scenario 2*: The Barcau river bed with dykes and the Salard non-permanent storage reservoir.

In this case the Salard temporary storage reservoir was added to the preceding scenario. Calculations show that in the Salard section the maximum flow is reduced from 450 m³/s to 375 m³/s. In the Berettyoujfalu section the corresponding discharge would be 367 m³/s. The maximum volume accumulated in the temporary storage reservoir is 4.36 million m³. In this case the dykes are not overtopped.

- *Scenario 3*: The Barcau river bed with dykes, the Salard temporary storage reservoir and the completed Marghita temporary storage reservoir (currently under construction).

In this case the Marghita temporary storage reservoir was added to the preceding scenario. The calculations show that the Salard temporary storage reservoir attenuates the maximum discharge from 420 m³/s to 361 m³/s. In the border section the maximum discharge would be 358 m³/s and 351 m³/s at Berettyoujfalu.

- *Scenario 4*: The Barcau river bed with dykes designed according to Class IV and the Salard temporary storage reservoir.

Taking into account that the dykes are designed according to Class IV (5% probability discharge), in the case of a 1% probability flood, the existing dykes would be overtopped and in the Salard reach the discharge would be almost equal to the one in natural conditions. In this case the Salard temporary storage reservoir would attenuate the maximum discharge from 370 m³/s to 336 m³/s. In the border section the maximum discharge would be 335 m³/s and 331 m³/s at Berettyoujfalu.

4. Conclusions

After analysing the above mentioned data it can be shown that the Salard temporary storage reservoir reduces the maximum flood corresponding to 1% frequency ($Q_{1\%}$) from 450 m³/s to 375 m³/s. Consequently, the maximum attenuated discharge is almost equal to that with a 1% discharge probability under the natural conditions and the water level at the border is 1 m lower than in the case without the temporary storage reservoir.

Presently the dykes in the Almasu Mic – Salard reach provide flood protection corresponding to the 5% probability level, and they would be overtopped in the case of a 1% probability flood (Scenario 1). In the Salard reach the maximum discharge would be almost equal to the one corresponding to the natural conditions; the temporary storage reservoir would reduce the discharge from 370 m^3/s to 335 m^3/s, and in the border section the level would be 1.20 m lower than in the case when the dykes in the Salard – Almasu Mic reach are not overtopped.

For flow up to 160 m^3/s, the corresponding water levels would be about 80 cm higher in 2004 than in 1974 and 1980. This increase was explained by the modification of the Manning coefficients. Changes in the rating curves and floods in the study area are presented in Figure 2.

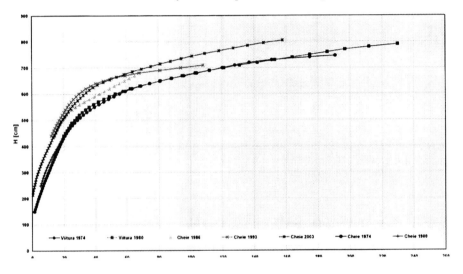

Fig. 2. The evolution of the rating curves at the Salard gauging station

References

Mazilu P, Mustatea A (2002) Studiu privind efectul acumularii nepermanente Salard asupra sectorului de rau de interes comun pentru Romania si Ungaria, intre statiile hidrometrice Salard si Berettyoujfalu (in Romanian, Study regarding the effect of the Salad temporary storage reservoir on the reach of common interest to Romania and Hungary between the Salard and Berettyoujfalu hydrometrical stations). S.C. AQUAPROIECT, Bucharest, Romania

Streng O, Gale M, Dima S, Crisan S, Craciun I, Craciun M, Kutz I, Paul T, Zbarcea F, Sidea I (2001) Hydrometrical Studies for the Oradea Hydrological Station (1990-2001). Crisul Water Authority, Oradea, Romania

FLOOD CONTROL MANAGEMENT WITH SPECIAL REFERENCE TO EMERGENCY RESERVOIRS

ZOLTÁN GALBÁTS[1]
*Körös Region Environmental and Water Directorate,
Városház u. 26. H-5700 Gyula, Hungary*

Abstract. The flood control system of the Körös rivers consists of four elements: main dykes, local dykes, emergency reservoirs and circle dykes of settlements. The heart of the flood control system are main dykes serving direct protection functions during every flood. Local dykes serve as back-ups and control the extent of inundation when the main dykes are breached. Circle dykes protect settlements built in the inundation area on flood plains. Emergency reservoirs or detention basins are facilities, which are suitable for temporary storage of flood water. They are activated only in extreme situations of threats to the main dykes to avert greater damages and flood disasters.

Keywords: flood control, emergency reservoirs, flood detention basins

1. Introduction

After the experience of the last decades of the 20th century, it became obvious that the traditional way of flood protection development by gradual strengthening and rising of dykes is not sufficient for the flood defence of the Körös region. The use of flood detention basins or emergency reservoirs offers an additional type of pro-active measures to strengthen the system of flood protection in the Körös system. Such reservoirs provide additional

---

[1] To whom correspondence should be addressed. Zoltán Galbáts, Körös Region Environmental and Water Directorate, Városház u. 26. H-5700 Gyula, Hungary; e-mail: galbats.zoltan@korkovizig.hu

flood water storage and help protect large portions of former floodplains on the protected side of the dyke system.

2. Drainage area of the Körös Rivers and the flood control system

Körös-valley District Environment and Water Directorate is situated in the south-east part of Hungary. The catchment area of the Körös and Berettyó Rivers covers 27,537 km² and the total length of rivers is 1,100 km. Forty-five percent of the region are hills and mountains (the highest peak is 1848 m), and 55% is lowland. The operating territory of the directorate is 4,108 km², with 2,822 km² (69%) being flood plains. Because of the shape of the catchment and the great differences in elevations, floods arrive to the plain extremely quickly (24-36 hours after the rainfall occurs). At some gauging station water level can increase even six-seven metres within 48 hours (see Fig. 1).

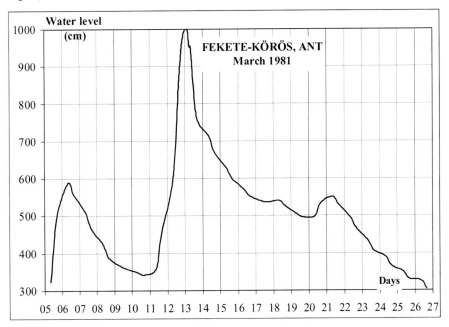

Fig. 1. Flood waves of the Fekete-Körös River

Consequently there is a real danger of breaching the dykes even at present. Peak flood levels are continually rising; on the Fekete-Körös river at Ant, between 1970-1981, by 0.92 m; at Remete, between 1966-1974, by 1.28 m; on the Fehér-Körös at Gyula, between 1962- 1974, by 1.11 m; and, on the Kettős-Körös at Békés, between 1966- 1974, by 1.31 m. The trend of increasing flood levels has not changed in recent years (Fig. 2). The aim of flood management is to convey high waters below the design flood levels even in extreme situations.

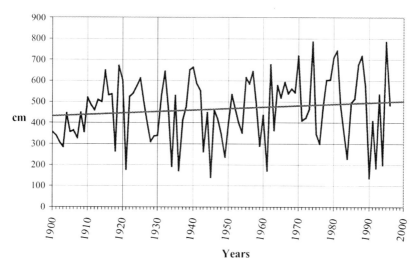

Fig. 2. Trend of increasing high flood levels

As a result of the development of the flood protection infrastructure since 1966 the flood control system of the Körös rivers consists of four parts: main dykes, local dykes, emergency reservoirs and circle dykes of settlements. The heart of the flood control system is the system of main dykes serving direct protection functions during every flood. Local dykes serve as back-ups and control the extent of inundation when the main dykes are breached. Circle dykes protect settlements built in the inundation area on flood plains. Emergency reservoirs or detention basins are facilities, which are suitable for temporary storage of flood water. They are activated only in extreme situations of threats to the main dykes for the sake of avoiding greater damages and flood disasters. The land inside the reservoirs is mainly used for agriculture and forestry. The main purpose of emergency storage is to temporarily store some quantity of flood waters and thereby reduce the flood crest.

3. Emergency reservoirs on the Körös Rivers

There are six emergency reservoirs on the Körös Rivers and the Berettyó in Hungary and three of them are on the territory of the Körös Region Directorate. After the extreme flood in the Tisza-valley in 1970, two emergency reservoirs were built in the region of the Directorate (in 1975-76 the Mérges reservoir and in 1975-77 the Mályvád resrevoir). As a consequence of the record flood levels observed on the Fehér-Körös in late 1995 – early 1996, the Kisdelta emergency reservoir was built in from 1997 to 1999.

Fig. 3. Emergency reservoirs on the Körös Rivers

3.1. MÉRGES EMERGENCY RESERVOIR

The Mérges emergency reservoir is located at the confluence of the Kettős-, and Sebes-Körös rivers. It is fitted with two 70 m long opening structures, built into the dykes of the Sebes-, and Kettős-Körös rivers, and activated by a controlled explosion removing the upper part of the dyke and thereby creating a side weir allowing flood waters to drain into the reservoir. The water temporarily detained in the reservoir can be released back to the Kettős-Körös river (after the stage subsides) at the opening site by two sluice-controlled culverts of $\varnothing$ 2x2.20 metres with a capacity 23.1 m^3/s.

The whole reservoir area is 18.23 km² and its storage capacity is 87.2 million m³. It was operated on two occasions; first from July 28 to Sept. 28, 1980, when 50 million m³ of water was stored with the surface level of 87.05 metres above the Baltic Sea Datum. The second time it was in use from Dec. 29, 1995 to May 31, 1996, when it was also opened from the side of the Kettős-Körös River; 31.8 million m³ of water was stored, and the water level reached 86.14 metres above the Baltic Sea Datum.

3.2. MÁLYVÁD EMERGENCY RESERVOIR

The Mályvád emergency reservoir is located on the left bank of the Fekete-Körös River. Main hydraulic structures were built into the dyke of the river. There is an opening site with a fixed threshold; the earthen dyke above the threshold structure can be removed by explosion. Explosives are inserted into tubes buried in the dyke. Stored water can be returned to the river through the opening site fitted with a 3x2.40x1.80 m sluice.

Most of the area is oak forest; game stock is very significant, particularly fallow deer. Thus, fourteen animal protection hills were built, each measuring 30x70 metres. The total area of this reservoir is 36.89 km² and the storage capacity is 75.0 million m³.

The reservoirs was operated on three occasions: Aug 1 – Sept. 6, 1980, when 19.0 million m³ of water was stored and the inundated area was 2.5 km²; Mar. 13 – Apr. 6, 1981, when the entire area of the reservoir was inundated, the stored volume was 75.0 million m³ and the water level was 90.12 m above the Baltic Sea Datum; and Dec. 29, 1995 – Jan. 22, 1996, when 7.4 million m³ of water was stored, 1.2 km² was inundated, and the water level was 87.55 m.

3.3. KISDELTA EMERGENCY RESERVOIR

The Kisdelta emergency reservoir is located at the confluence of the Fehér- and Fekete-Körös rivers. In the Fehér-Körös river dyke, there is an opening site with a fixed threshold; the earthen dyke above the threshold structure can be removed by explosion. Explosives are inserted into tubes buried in the dyke. Stored water can be returned to the river through the opening site and a sluice with dimensions 2x1.50x1.50 m on the Itceéri channel. 14 million m³ (53.8 %) of the stored water can be released only by the pumping station Gyula II with the highest capacity of 3.8 m³/s. The area of the reservoir is 5.80 km² and the storage capacity is 26.0 million m³. There was no need to operate the reservoir as yet, even though some preparations for opening were initiated in 2000.

4. Conditions for use of reservoirs and the related decision making

4.1. OPERATION OF EMERGENCY RESERVOIRS

Emergency reservoirs can be used in four essentially different situations, and flood control practice in the Körös region shows examples of all of them:

- to reduce flood levels higher than the flood control structures can bear (Mályvád 1981);
- defence against ice floods threatening to breach dykes (Berettyó 1966);
- to avoid flood disasters in the case of dangerous weakening of main dykes caused by a long-term hydrostatic load (Berettyó 1970); and,
- as a follow up action to reduce adverse impacts of a dyke breach (Mályvád 1980, Mérges 1980).

The effect of an emergency reservoir operation on the main river flow depends on the location:

- downstream from the opening site on the river, water withdrawal dominates
- upstream on the river appears an intensive draw-down of water, and
- on tributaries, a moderate impact on the fall of water level is observed.

Experiences of previous operations of emergency reservoirs indicate that the results to be expected have to be evaluated carefully in advance for all typical sections of the river system. This is especially important for the Körös Rivers, with a number of tributaries, where six emergency reservoirs can be operated simultaneously in different combinations.

4.2. OPTIMISATION OF OPERATIONS

Optimisation of the operation of emergency reservoirs in a river-system can be carried out in three major groups of actions and eleven steps as described below.

4.2.1. *Forecasting of hydrological situation*

1. The existing hydrological condition of the river-system is characterised by continuously incoming information. Calculations of actions for preparing the emergency storage activation start when data or information received signal a critical change of the river system from the point of view of flood defence.

2. Primary evaluation of conditions by hydrological and hydrometeorological analyses as well as by using a flood forecasting system.

3. The flood forecasting is based on data from the main gauging stations of the Körös Rivers which are complemented by data from the meteorological satellite receiver system and the precipitation radar. These data are used in such a way, as to ensure the greatest possible lead time as well as to produce the most accurate water level forecasts for defining the defence strategy and organising flood defence measures. With respect to the forecasting, the gauging stations in Romania are naturally the most important ones. In the case of flood emergency, operational data on water level and discharge measurements arrive from sixteen stations five times per day, as well as the data concerning the time and water level pertinent to the automatic activation of the emergency reservoirs Tamasda and Zerindul Mic.

4. If the calculations described in step 2 indicate that the expected water levels will surpass the flood control alarm level III, further calculations have to be made. Otherwise the calculation process returns back to step 1 again.

5. Repeated predictions should be made for water levels at standard stations of the river system, and then longitudinal sections of the culminating water levels have to be established.

6. The calculated flood crest elevations should be compared with the physical parameters of flood protection structures (e.g., the design elevation, height of dykes).

7. On the basis of data and results of calculations and other available information, flood control managers have to make a conceptual decision about the operation of emergency reservoirs. Calculations have to be continued if the decision to operate the reservoir was made. The purpose of such calculations is to establish the plan of action. Otherwise the calculation process returns back to step 1 again.

8. Definition of possible options and evaluation of the impact of reservoir operation.

9. Operation of one or more emergency reservoirs in the Körös valley should be taken into consideration and the feasible combinations have to be also examined.

10. Efficiency of one or more options has to be determined. Order of operational steps and their mutual impacts, as well as the processes

occurring inside the reservoirs should be examined. Predictions of the impacts of different options need to be made.

11. Based on the above mentioned calculations the final selection of operational steps has to be made from the investigated options.

4.2.2. *Action plan for operation of emergency reservoirs*

Based on the selected option and the contingency plan (produced in 2001) we should prepare an action plan for the execution of operation. The plan among others should define the required starting time of preliminary works with respect to the optimal time of reservoir opening.

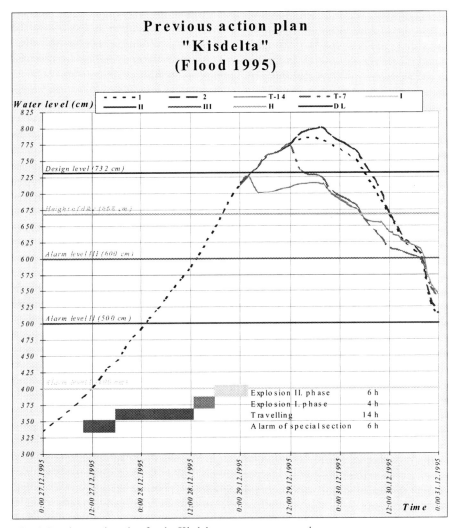

Fig. 4. Previous action plan for the Kisdelta emergency reservoir

The activation of an emergency reservoir requires a number of preparatory steps, which take significant time durations, compared to the available lead time of prediction. Thus lines of actions have to be set to start when information for decision making in not yet complete. Actions of preparation should be organised and executed so that all the tasks will be always executed in a standard time. This does not mean that reservoir operation has to be fully completed, because the preparation process can be stopped. Such a situation occurred in April, 2000, when the extreme flood on the Fehér-Körös River triggered the operation of the Kisdelta emergency reservoir. Then the preparatory works were stopped because flood levels started to recede as a result of an upstream dyke breach.

The critical path analysis of preparatory works for emergency reservoirs in the Körös- valley produced estimates of the required time as about 24-26 hours.

The development of a software tool to support flood defence operations related to emergency reservoirs in the Körös valley has been commissioned.

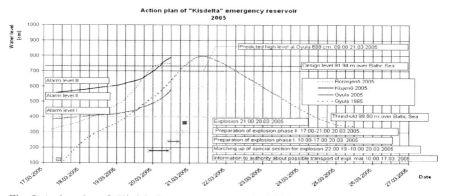

Fig. 5. Action plan of "Kisdelta" emergency reservoir (2005)

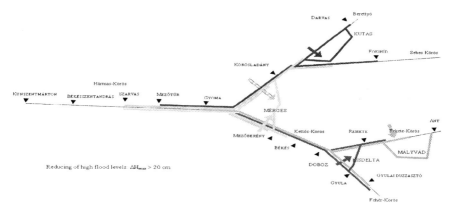

Fig. 6. Interactions of the emergency reservoirs

5. Execution of activation of emergency storage

When the possibility arises that on the basis of hydrologic predictions and calculations, the emergency reservoir operation may become necessary, the Directorate (with the consent of the Békés County Defence Committee) forwards this information to the National Technical Directing Group (NTDG) and requests mobilisation of a special team for blasting the reservoir opening. The permission for operation is granted by the Government Commissioner designated at the national level.

Preliminary works include controlling hydraulic structures and dykes, advising agricultural land users, railway and forestry authorities, and organising the protection of opening sites.

Actual activation operation consists of two parts; the first part takes about seven hours and can be stopped any time. The second part needs about three hours, which covers building a network of explosives, and is irreversible. The execution of explosion can be started by a written order of the Director.

Filling up the area of the reservoir with water should be observed from air or space. Releasing water back to the river has to be started as soon as possible to ensure free storage capacity in anticipation of a potential next flood wave. When the water levels fall below the mark of the opening site, reconstruction works have to be started.

6. Conclusions

Based on the experience with operation of emergency reservoirs, the following can be concluded:

- The reservoir filling structure has to release flood waters quickly, and be designed as an appropriately reinforced threshold structure (with a removable embankment on the top) to avoid overfilling the reservoir and prevent destruction of the structure when returning stored water back to the river.
- For dewatering of the entire area of the reservoir, special hydraulic structures (outlets) should be built. It is practical to determine the threshold of the outlet (controlled by a sluice) so that water needed for ecological requirements and irrigation can be released. The outlet and the sluice should be designed to facilitate dewatering in not more than thirty days.
- To protect wildlife, protection hills should be established where necessary.

- Operation of an emergency reservoir significantly affects downstream sections of the river; it is considered to be significant if reduction of flood peaks at the downstream reach is more than twenty centimetres.
- To achieve the greatest benefit of emergency reservoirs, they have to be opened at the right time. The essential condition for achieving this is the accurate prediction of flood hydrograph, which requires a telemetry system extending across the entire river basin.
- To determine the most favourable time of reservoir activation, the use of hydrodynamic models for flood data processing is required.

Operation of emergency reservoirs needs very careful considerations, with special attention paid to direct and indirect damages and the costs of reconstruction, which make the operation of emergency reservoirs very expensive.

After the experience of the last decades at the end of the 20th century it became obvious that the traditional way of flood protection development by the gradual strengthening and rising of dykes is not sufficient for the flood defence in the Körös region. The use of flood detention basins or emergency reservoirs offers an additional type of pro-active measures to strengthen the system of flood protection in the Körös system.

STRUCTURAL FLOOD CONTROL MEASURES IN THE CRISUL REPEDE BASIN AND THEIR EFFECTS IN ROMANIA AND HUNGARY

MITRUȚ TENTIȘ[1], MARINELA GALE, CEZAR MORAR
Oradea's Crisuri Rivers Authority, Romania National Water

Abstract. Flood protection by structural measures has been practiced in the Crisul Repede basin for more than 130 years. Some of such developments, particularly those implemented during the last several decades, are reviewed and their effects on flood peak attenuation are demonstrated for the flood of Dec. 1995-Jan. 1996. Reconstruction of flood hydrographs for two cases, with and without reservoir storage in place, indicates a significant flood peak reduction on both the Romanian and Hungarian territories.

Keywords: the Crisul Repede River, transboundary floods, flood control, flood storage reservoirs, attenuation of flood peaks by storage,

1. Introduction

The Crisuri Rivers Basin (Fig. 1), which is located in north-western Romania and eastern Hungary, has a surface area of 27.500 km² (about 14,860 km² on the Romanian side, or 54%). The main basin rivers, the Barcau, Crisul Repede, Crisul Negru and Crisul Alb cross the border, merging two by two into Double Crisuri, and then into Triple Crisuri, and finally they discharge into the Tisa river near Csongrad. As a principal characteristic of the basin, one can identify the contrast between high flows

---

[1] To whom all correspondence should be addressed. Mitruț Tentiș, Oradea's Crisuri Rivers Authority, 35 Ion Bogdan Street, 410 125 Oradea, Bihor, Romania; e-mail: mitrut.tentis@dac.rowater.ro

in the mountain region, where these rivers originate, and low flows in the Western Plain.

The Crisul Repede River Basin is located in the central part of the Crisuri Rivers Basin and has a surface area of 9,211 km^2 (5,963 km^2 on the Romanian territory). The Crisul Repede River (209 km long) crosses different forms of relief occurring in several steps, starting with the mountainous region in the East (with the maximum elevation of 1836 m), followed by hills and the valley region, and finally the downstream plains positioned just about 100 m above the sea level (see Fig. 2).

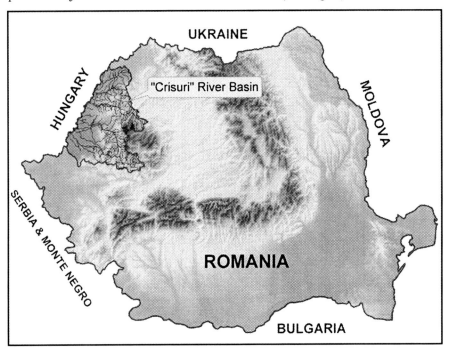

Fig. 1. Map of the Crisuri Rivers basin

The prevailing atmospheric circulation is in the westerly direction; the air-masses coming into contact with the mountains can produce high precipitation in short periods of time and cause floods with quick propagation to the downstream areas.

The basin shape is asymmetrical; the most powerful tributaries (the Calata, Hent, Dragan and Iad) come from the mountainous region and represent an important hydroelectric potential. In the valleys and the plain region, the river has mild slopes and strongly meanders, which contributes to flooding. Flow characteristics at key stream-gauging stations are listed in Table 1.

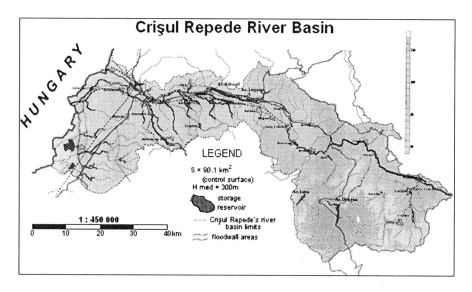

Fig. 2. Map of the Crisuri Repede River basin

Table 1. Flow characteristics at the key stream-gauging stations

No.	River	Gauge Station	A (km^2)	Stage (m)	Q_0 (m^3/s)	Q_{max} 1% (m^3/s)	Q_{min} 95% (m^3/s)
1	Crişul Repede	Ciucea	814	904	12.1	660	1.10
2	Crişul Repede	Vadu Crişului	1 328	821	20.4	820	2.00
3	Crişul Repede	Oradea	2 176	630	25.4	1, 000	2.80

The hydrologic and hydraulic potential of the Crisul Repede River was addressed for the first time in the 1870s by initiating complex hydro-technical planning. A large number of dams, dykes, and other hydraulic structures, were built to support important economic activities and inevitably caused changes in the flow regime and damaged the ecological functions and biodiversity of the river system. These factors, along with changes in land use throughout the catchment, have caused significant environmental damage, such as increased erosion, reduced sediment transport, and reduced self-purification capacity of the rivers.

floods water levels either positively or negatively. The retention capacity of the flood plains changes the water levels and flood wave propagation time, and consequently influences the times of arrival of flood waves from the main river reaches and tributaries. Anthropogenic influences on watercourses may have beneficial or negative consequences; beneficial effects in small sub-catchments may have adverse consequences for the main river reaches.

While the large reservoirs constructed on rivers within the Crisul Repede River catchment do have some flood attenuation effects, those may have been counteracted by changes in the natural storage conditions due to: (a) urbanisation (the economic development, especially in the city of Oradea, imposed additional demands for industrial water supply), (b) changing land use (changing vegetation, depressions, soil water holding capacity), (c) deforestation, and (d) reduction of natural flood plains.

To meet the water supply demands, a reservoir was built on the Crisul Repede River and a water intake with a capacity of 9 m^3/s started working in 1962. Because of flow fluctuation in the Crisul Repede from 1 to 600 m^3/s, sometimes there is not enough water to meet the demand.

To solve this problem the Lesu Reservoir in Iada Valley (total volume of 31.7 mil m^3) was brought into operation in 1973 (Fig. 3). This multi-purpose reservoir was built to secure water supply needs, provide hydroelectric power, and optimise flood protection. The catchment of the Lesu Reservoir has a surface area of 90.1 km^2, an elevation of 993 m (related to the Black Sea level) and is fed by a number of streams. Other reservoir data are as follows: Dam type - riprap with a concrete apron, dam height and length - 60.5 and 180 m, respectively, detention storage - 25.5 mil m^3, and hydroelectric power plant potential - 3.4 MW.

Fig. 3. Leşu Reservoir

The main characteristics of three major reservoirs in the Crisul Alb Basin are listed in Table 2.

Table 2. Characteristics of three reservoirs in the Crisul Alb Basin (Drăgan, Lugas and Tileagd)

Reservoir Characteristic	Drăgan	Lugas	Tileagd
Dam type	Concrete double arch	Earthen dam with concrete apron	Earthen dam with concrete apron
Catchment area (km^2)	156	1,764	1,831
Dam height (m)	120	37.5	37.5
Dam length (m)	424	350	500
Storage volume (mil m^3)	124	74.5	63.3
Normal retention volume	112	63.5	53.2
Minimum exploitation volume	14.6	14.7	10.1
Year of completion	1988	1988	1988

To improve the hydroelectric potential and flood protection, a whole cascade of reservoirs with 6 hydroelectric plants was completed after 1980. The installed hydroelectric generation capacity is 213 MW and the total storage volume is 293.5 mil m^3. Completion of this reservoir system in 1988 greatly reduced the flood risk in the Crisul Repede Basin.

During the period from Dec. 1995 to Jan. 1996, an extreme flood occurred throughout the Crisuri Basin. It was caused by a dramatic weather warming, with temperatures over 15^{0}C that produced a rapid melting of the snowpack, more than 1 m deep, in the mountains. Large floods occurred especially on the Crisul Negru River and on the Crisul Alb River. At the Gyula (Hungary) Hydrometrical Station a record stage of all times was observed.

On the Crisul Repede River the complex hydrotechnical structures managed to attenuate the flash flood. Altogether, the four important reservoirs accumulated about 70 mil m^3 of water from Dec. 23 to 27, 1995 (Figs. 4-7).

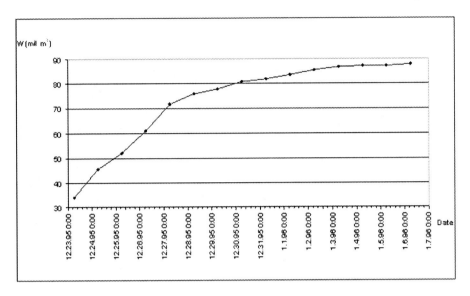

Fig. 4. Hydrograph of the Dragan Reservoir [W = f (t)]

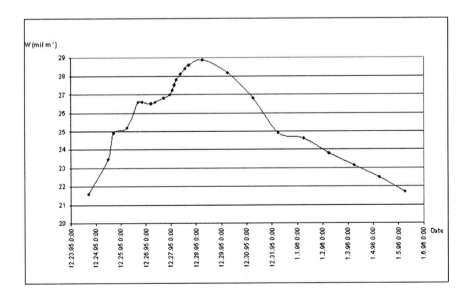

Fig. 5. Hydrograph of the Lesu reservoir [W = f (t)]

FLOOD CONTROL IN THE CRISUL REPEDE RIVER BASIN 283

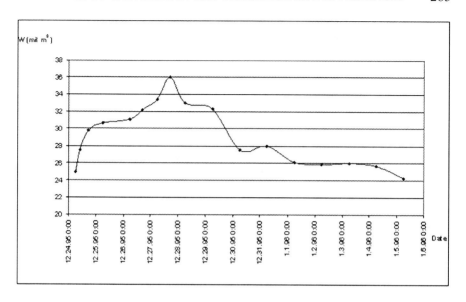

Fig. 6. Hydrograph of the Lugas reservoir [W = f (t)]

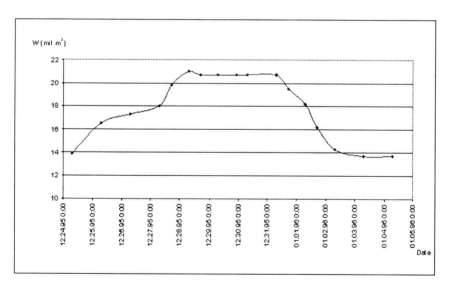

Fig. 7. Hydrograph of the Tileag reservoir [W = f (t)]

On Dec. 27, at the Oradea Hydrometrical Station, located downstream from the last reservoir, the maximum flow was 346 m^3/s. The maximum reconstituted flow was 637 m^3/s (without reservoirs), with a certainty of 5% (see Fig. 8).

The reduction of the maximum flow on the Crisul Repede River in the border region of almost 300 m³/s had positive effects in downstream reaches on both Romanian and Hungarian territories. Because of the reduced maximum flow on the Crisul Repede River at the confluence with the Double Crisuri, the floods effects on the Triple Crisuri, all the way to the confluence with the Tisa, were also considerably reduced.

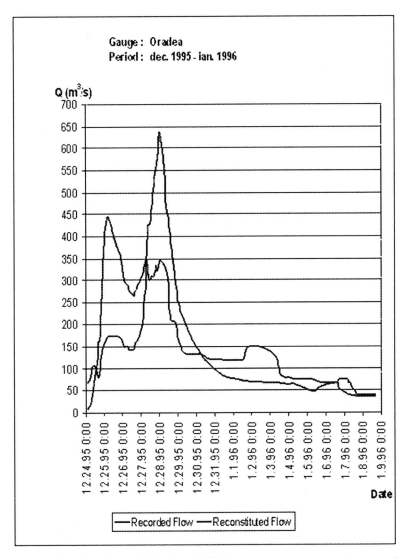

Fig. 8. Oradea flash flood (Dec. 1995 – Jan. 1996) (the upper curve represents the reconstituted hydrograph)

2. Conclusions

The structural and non-structural flood protection measures in the upper Crisuri Rivers Basin contribute to flood mitigation and reduced flood damages in both Romania and Hungary. All Romanian hydrotechnical projects currently under development, including the Mihaileni retention reservoir (10.3 mil m^3 volume), temporary storage reservoirs Sicula and Zarand (12 mil m^3 total volume) on the Crisul Alb River, temporary storage reservoir Ginta (14 mil m^3) on the Crisul Negru River, and the permanent reservoir Suplac (15.4 mil m^3) on the Barcau river, will all contribute to better flood management and mitigation of flood effects.

CONTRIBUTION OF EARTH OBSERVATION DATA SUPPLIED BY THE NEW SATELLITE SENSORS TO FLOOD MANAGEMENT

GHEORGHE STANCALIE[1], SIMONA CATANA, ANISOARA IRIMESCU, ELENA SAVIN, ANDREI DIAMANDI, ALINA HOFNAR, SIMONA OANCEA
National Meteorological Administration (NMA), Bucharest, ROMANIA 031686

Abstract. The risk of flooding due to runoff is a major concern in many areas around the globe and especially in Romania. In recent years river floods and associated landslides have occurred quite frequently in Romania, some affecting small isolated areas, but others affecting large areas of the national territory. The main objective of the NATO SfP project "Monitoring of extreme flood events in Romania and Hungary using EO data" is to improve the existing local operational flood hazard assessment and monitoring, using the functional facilities supplied by the GIS info-layers, in combination with Earth Observation (EO) data-derived information, Digital Elevation Models (DEM) and hydrological modelling. The study area is situated in the Crisul Alb - Crisul Negru - Kőrös transboundary basin, crossing the Romanian – Hungarian border. Orbital remote sensing can provide the information necessary for flood hazard and vulnerability assessment and mapping, which are directly used in the decision-making process. The EO data-derived information of the land cover/land use is important because it allows periodical updating and comparisons, and thus contributes to characterisation of the human presence and the vulnerability, as well as to the evaluation of the impact of the floods. In order to obtain high-level thematic products the data extracted from the satellite images must be integrated with other geo-information data (topographical, pedological, meteorological data) and hydrologic/hydraulic model outputs. The paper presents the specific methods developed for deriving satellite-based applications and products useful for flood disaster

---

[1] To whom correspondence should be addressed. Gheorghe Stancalie, National Meteorological Administration, Remote Sensing & GIS Laboratory, 97, Soseaua Bucuresti-Ploiesti, Sector 1, 013686 Bucharest, Romania; e-mail: gheorghe.stancalie@meteo.inmh.ro

assessment and hazard reduction. Using the optical and microwave data supplied by the new satellite sensors (U.S. DMSP/Quikscat, RADARSAT, LANDSAT–7/TM, EOS-AM "TERRA"/MODIS and ASTER) different products like accurate, updated digital maps of the hydrographical network and land cover/land use, mask of flooded areas, multi-temporal maps of the flood dynamics, hazard maps with the extent of the flooded areas and the affected zones, etc. have been obtained. These results, at different spatial scales, include synthesised maps, easy to access and interpret, and that can be combined with other information layers from the GIS database and can accept rainfall-runoff model outputs. The presented applications will contribute to preventive consideration of the extreme flood events by planning more judiciously land-use development, by elaborating plans for food mitigation, including infrastructure construction in the flood-prone areas, and by optimisation of the flood-related spatial information distribution to the end-users.

Keywords: flood risk, flood management, EO data, GIS, the Crisul Rivers, the Körös Rivers, Transboundary flooding

1. Introduction

Flooding remains to be the most widely distributed natural hazard in Europe leading to significant economic and social impacts. In the assessment and analysis of flood risk it is important to remember that risk is entirely a human issue. Floods are a part of the natural hydrological cycle and occur randomly. The risk arises because the human use river flood plains which conflicts with their natural function of conveyance of water and sediment.

Orbital remote sensing of the Earth can make fundamental contributions towards reducing the detrimental effects of extreme floods. Effective flood warning requires frequent radar observations of the Earth's surface through the cloud cover. In contrast, both optical and radar wavelengths will increasingly be used for disaster assessment and hazard reduction. These latter tasks are accomplished, in part, by accurate mapping of flooded lands, which is commonly done over periods of several or more days. The detection of new flood events and public flood warning is still experimental, but evolves rapidly; radar sensors are preferred due to their cloud penetrating capability as reported by Brakenridge et al. (2003). Relatively low spatial resolution, but wide-area and frequent coverage, are appropriate; the objective is to locate where within a region or watershed the flooding occurs, rather than to map the actual inundated areas. The rapid-response flood mapping provides information useful for disaster assessment, and has become a common activity in many countries.

In recent years river floods and associated landslides have occurred quite frequently in Romania, some affecting small isolated areas, but others affecting large areas of the national territory. One region, which suffers from flood damages on a regular basis, is the transboundary area of the Crisul Alb and Crisul Negro rivers flowing from Romania into Hungary, where they are known as the Kőrös rivers. Floods in this area start in the mountainous terrain of the upper parts of the basin in Romania and propagate to the plains in Hungary.

Historically, there has been a close co-operation between both countries in flood management in this area. The issues connected with the Transboundary Rivers crossing the Romanian – Hungarian border are covered by the bilateral Agreement for the settlement of the hydrotechnical problems, which was issued in 1986. To facilitate the implementation of this agreement, working groups from the Crisuri Water Authorities in Oradea, Romania and Körös Valley District Water Authority (KOVIZIG) in Gyula, Hungary meet regularly to address the issues of mutual interest (Brakenridge et al. 2001).

The flood forecast and defence related information provided by Romania to Hungary is presently based entirely upon the ground-observed data, which are mostly collected by non-automatic hydrometeorological stations. Such data are somewhat limited in terms of spatial distribution, temporal detail, and speed of collection and transmission, and these limitations should be remedied.

Recognising the threat of floods and the need for improved flood management in this area, at the initiative of the Romanian Meteorological Administration, an international team, with representatives of Hungary, Romania and USA, proposed a project on "Monitoring of Extreme Flood Events in Romania and Hungary Using EO (Earth Observation) Data" to the NATO Science for Peace (SfP) Programme. The project aims to provide to the local and river authorities as well as to other key organizations an efficient and powerful flood-monitoring tool, which is expected to significantly improve the efficiency and effectiveness of the action plans for flood defence. Some applications, to flood hazard and vulnerability assessment and mapping, developed in the framework of this NATO SfP project are presented.

2. Study area

The study area represents the Crisul Alb/Negru/Kőrös transboundary basin spanning across the Romanian-Hungarian border, with a total area of 26,600 km^2 (14,900 km^2 on the Romanian territory). In Romania, the catchment (basin) comprises mountainous areas (38%), hilly areas (20%) and plains (42%). About 30% of the catchment is forested. On the Hungarian side, the catchment relief represents plains (Fig. 1). Annual precipitation ranges from 600-800 mm/year in the plain and plateau areas to over 1200 mm/year in the mountainous areas of Romania. This precipitation distribution can be explained by the fact that humid air masses brought by fronts from the Icelandic Low frequently enter this area. The orography of the area (Apuseni Mountains) amplifies the precipitation on the western side of the mountain range. Thus, the Crisuri Rivers Basin frequently experiences large precipitation amounts in short time intervals and the frequency of such events seem to be increasing in recent years.

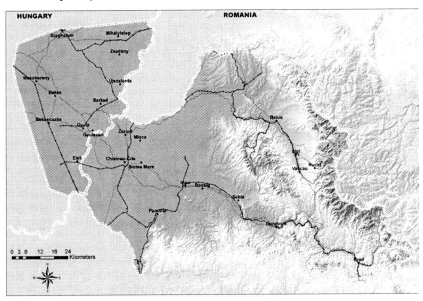

Fig. 1. Study area: the Crisul Alb - Crisul Negru - Körös transboundary basin, crossing the Romanian – Hungarian border

In terms of streamflow formation, there is a marked difference between high rates of mountain runoff and low rates of runoff in plains. Thus, runoff flood waves formed quickly in the Romanian part of the basin move rapidly to the plains in the Hungarian part of the basin, which is characterised by relatively slow flows and a potential for inundation. In terms of flood forecasting, the Romanian part of the basin is of greater interest with respect to flood formation, which is also reflected in this paper. The hydrometry of the study area is well established. There are 62 hydrometric stations in Romania, on the Crisul Alb and Negru (and their tributaries); 7 of these stations have flow records longer than 80 years. On the Hungarian territory, the hydrometric stations at Gyula and Sarkad are particularly of interest. In Gyula, the flow was decreasing in time, but the stage was rising, in part due to the hydrotechnical structures; at Sarkad, both discharge and stage were increasing, which reflects more natural conditions, without much change in the river channel geometry.

The list of significant floods includes the events of June 1974, July-August 1980, March 1981 and December 1995-January 1996, March 2000, April 2000 and April 2001. The spring 2000 flood caused on the Romanian territory damages of more than $US 20 million, including damages to houses, roads and railways, bridges, hydraulic structures; loss of domestic animals, and business losses. On the Hungarian territory, a particularly notable was the flood of summer 1980, with total losses of $US 15 million, including destruction of farmhouses and large losses in agriculture.

The frequency and importance of floods in the study region require further work to reduce flood damages and improve flood monitoring by the agencies in charge of flood protection, such as government agencies, civil protection authorities and municipalities. To mitigate flood impacts in the study area, structural and non-structural measures have been undertaken in the past. The Romanian area is defended by dykes along the Crisul Alb River and the Crisul Negru River. These dykes were built in the 19th century for a 20-year design return period and were further improved in later years. Currently, the dykes on the right bank of the Crisul Negru River and the Teuz River (43 km) are designed for a 50-year return period, and on the Crisul Alb, 67 km of dykes on the right bank and 59 km on the left bank are designed for a 100-year return period. In spite of these improvements, in April 2000, the right bank dyke of the Crisul Negru broke near the village Tipari (a 130 m wide breach) and caused significant flooding of, and damages in, the adjacent territory. Other structural flood protection measures include permanent retention storage facilities (total volume of 34 x 10^6 m^3) and temporary storage facilities (a total storage volume of almost 80 x 10^6 m^3).

On the Hungarian side, in the Kőrös valley, high flood potential is recognised and exacerbated by low flood plains. Much of the area is, therefore, protected by flood dykes. More than 440 km of dykes are maintained by the Kőrös River Authority (KOVIZIG). Following the 1979 flood, construction of detention (emergency) reservoirs started. Altogether, these reservoirs provide storage capacity of 188 million m^3 and serve to reduce critical flood levels. The reservoirs are activated during floods, by a controlled explosion opening a protected spillway (a side weir) in flood dykes. Detained water inundates areas with lower intensity of agricultural activities and causes limited damages. Nevertheless, the reservoirs are activated only when necessary to avoid higher losses caused otherwise (Balint 2004).

The analysis and management of floods constitute the first indispensable step towards, and a rational basis for, the development of flood protection. Where certain flood risk levels are inevitable, the affected parties must know it and be appropriately warned. To reduce the frequency and magnitude of the damages due to flooding, comprehensive, realistic and integrated strategies must be developed and implemented.

The flood forecasting and monitoring systems existing in the study area do not reflect well the spatial distribution of floods and the related phenomena (pertaining to geographic distances or patterns) in both pre- and post-crisis phases. To mitigate these limitations, the SfP project was initiated with emphasis on a satellite-based surveillance system connected to a dedicated GIS database that will offer a much more comprehensive evaluation of the extreme flood effects. Also, so far, the flood potential, including the risk and the vulnerability of flood-prone areas, have not been yet quantitatively assessed. An inventory of the past floods observed by the EO facilities would allow a more cost-effective design of structural and non-structural measures for flood protection and disaster relief. Finally, such data also provide important validation of the hydrological modelling-based flood risk assessment, because they show the actual extent of past flooding.

3. EO data used

Apart from ground information on the occurrence and evolution of the flood, NOAA/AVHRR satellite data (locally received), microwave data from U.S. DMSP and Quikscat and follow-on satellites, supplied by the orbital platforms (SPOT, IRS, LANDSAT–7, RADARSAT, QUIKSCAT, EOS-AM "TERRA" and EOS–PM "AQUA"), substantially contribute to determining the flood-prone areas. The accuracy and level of detail of the geo-data produced by the high resolution images are especially useful for flood disaster management. Tables 1 and 2 provide a list of basic features of important space-borne remote sensing observation systems.

Table 1. Spectral and spatial characteristics of important high resolution satellite sensors

Satellite	Ground resolution - PAN (m)	Ground resolution - MS (m)	Scene size Swath width (km)	Spectral bands
LANDSAT - ETM	15	30	170 x183	VIS, NIR, TIR
ASTER	-	15	60 x 60	VIS, NIR, TIR
SPOT 2 & 4	10	20	60 x 60	VIS, NIR, SWIR
SPOT 5	2.5 & 5	10, 20	60 x 60	VIS, NIR, SWIR
IKONOS	1	4	11	VIS, NIR
QUICKBIRD	0.6	2.5	16.5	VIS, NIR
ORBVIEW 3	1	4	8 x 8	VIS, NIR

Table 2. Viewing characteristics of important high resolution satellite sensors

Satellite	Nadir revisit time (days)	Off-nadir revisit time (days)	Stereo
LANDSAT - ETM	16	No	No
ASTER	16	Variable	Yes/ in-track stereo
SPOT 2 & 4	26	1 – 3	Yes/ along track stereo
SPOT 5	26	1 – 3	Yes/ along track stereo
IKONOS	-	3	Yes/ in-track stereo+along track stereo
QUICKBIRD	1 – 3.5	-	Yes/ in-track stereo+along track stereo
ORBVIEW 3	15		Yes/ in-track stereo+along track stereo

RADARSAT-1 and 2, based on Synthetic Aperture Radar microwaves, penetrates clouds and rain and is very efficient for flood monitoring. RADARSAT operates in different modes with selective polarisation with a nominal swath width varying from 50 km (for the fine beam mode) to 500 km (for the ScanSAR wide beam mode). The approximate resolution varies between 10 x 9 m (for the fine beam mode) and 100 x 100 m (for the ScanSAR wide beam mode). RADARSAT data with finer resolution allow the structural flood damage assessment.

SeaWinds on QuikSCAT is a new space borne Ku-band scatterometer with the resolution of about 25 x 25 km, the swath of 1,800 km (for a vertical polarisation (VV) at a constant incidence angle of 54°) and of 1400 km (for a horizontal polarization (HH) at 46°). SeaWinds can provide nearly daily global coverage with the capability to see through clouds and darkness. It can, in principle, detect where flooding is occurring without necessarily imaging.

Especially the new American TERRA and AQUA platforms, equipped with different sensors such as MODIS and ASTER, can provide comprehensive series of flood event observations with much higher spatial resolution where available.

The Moderate Resolution Imaging Spectroradiometer (MODIS) continues the lineage of the Coastal Zone Colour Scanner (CZCS), the Advanced Very High Resolution Radiometer (AVHRR), the High Resolution Infrared Spectrometer (HIRS), and the Thematic Mapper (TM). MODIS has 36 spectral bands with centre wavelengths ranging from 0.412 mm to 14.235 mm. Two of the bands are imaged at a nominal resolution of 250 m at nadir, five bands are imaged at 500 m, and the remaining bands at 1000 m.

The Advanced Spaceborne Thermal Emission and Reflection Radiometer (ASTER), an imaging instrument, acquires 14 spectral bands and can be used to obtain detailed maps of surface. Each scene covers 60 x 60 km. The ASTER spectral bands are organised into three groups: 3 bands with 15 m resolution in VNIR, 6 bands with 30 m resolution in SWIR and 5 bands with 90 m resolution in TIR.

The information provided by these new sensors is of a higher quality than previously possible. Information obtained from these optical and radar images has been also used for the determination of certain parameters necessary to monitor flooding: hydrographic network, water storage, size of the flood-prone area, and land cover/land use features.

A Satellite Image Database (SID) containing data provided by different platforms and sensors has been set up. The purpose of the SID is to gather information about the raw satellite scenes available as well as of the derived products and make it available in a simple format. This information is useful to test the processing and analysis algorithms in order to establish an operational methodology for the detection, mapping and analysis of flooding. The SID was built in Microsoft Works and will be available on-line on the file server, being updated as new satellite images are acquired. Each record of the database describes the characteristics of each satellite image: platform, sensor, date and time of data acquisition, duration of pass, spectral band, coordinates of the area covered, projection, calibration, size, bits/pixel, image file format, physical location (machine, directory), origin of data, type (raw/processed), type of processing applied, algorithm used, quick-look available, and cloudiness.

4. Methods for obtaining useful products for flood risk assessment

It is generally recognised that the management and mitigation of flood risk require a holistic, structural set of activities, approached in practice on several fronts with appropriate institutional arrangements made to deliver the agreed standard services to the community at risk (Samuels 2004). The flood management includes activities conducted before, during and after the flood. Pre-flood activities are:

- Flood risk management for all causes of floods;
- Disaster contingency planning to establish evacuation routes, critical decision thresholds, public service and infrastructure requirements for emergency operations, etc;
- Construction of flood defence infrastructure, both physical defences and implementation of forecasting and warning systems;
- Maintenance of flood defence infrastructure;

CONTRIBUTION OF EO DATA TO FLOOD MANAGEMENT   295

- Land-use planning and management within the whole catchment;
- Discouragement of inappropriate development within flood plains; and,
- Public communication and education about flood risk and actions to take in a flood emergency.

Operational flood management includes:

- Detection of the likelihood of a flood formation;
- Forecasting of future river flow conditions from the hydrological and meteorological observations;
- Warning issued to the appropriate authorities and the public on the extent, severity and timing of the flood; and,
- Response to the emergency by the public and authorities.

Post-flood activities are:

- Relief for the immediate needs of those affected by the disaster;
- Reconstruction of damaged buildings, infrastructure and flood defence;
- Recovery of the environment and generation economic activities in the flooded areas; and,
- Review of the flood management activities to improve the process and planning for the future events in the area affected.

For flood surveying, optical and radar satellite images can provide up-to-date geographical information. Integrated within the GIS, flood derived and landscape descriptive information is helpful during characteristic phases of the flood (Tholey et al. 1997):

- Before flooding, the images serve to describe the land cover of the studied area under normal hydrological conditions;
- during flooding the image data provide information on the inundated zones, flooding extent, and flood evolution; and,
- after flooding, the satellite images point out the flood effects, showing the affected areas, sediment deposits and debris, with no information about the initial land cover description unless a comparison is performed with a normal land cover description map or with pre-flood data.

In order to obtain high-level thematic products, the data extracted from the EO images must be integrated with other non-space ancillary data (topographical, pedological, meteorological data) and hydrologic/hydraulic models outputs. This approach may be used in different phases of establishing the sensitive areas such as: the management of the database - built up from the ensemble of the spatially geo-referenced information; the elaboration of the risk indices from morphological, meteorological and hydrological data; the interfacing with models in order to improve their compatibility with input data; recovery of results and the possibility to work out scenarios; and, presentation of results as synthesised maps easy to access and interpret, and adequate for combination with other information layers from the GIS database.

The products useful for flooding risk analysis refer to: accurate updated maps of land cover/land use, comprehensive thematic maps at various spatial scales with the extent of the flooded areas and the affected zones, and maps of the hazard prone areas.

Optical and radar satellite data have been used to conduct inventories of different kinds of flood related thematic information.

A series of specific processing operations for the images were performed, using the ERDAS Imagine software: geometric correction and geo-referencing in the UTM or STEREO 70 map projection system, image improvement (contrast enhancing, slicking, selective contrast, combinations between spectral bands, re-sampling operation), and statistical analyses (for the characterisation of classes, the selection of the instructing samples, conceiving classifications).

Optical high-resolution data have been used to perform inventories under normal hydrological conditions as well as for determining the hydrographic network. The radiometric information contained in these images allows the derivation of both biophysical criteria and human activities, through supervised standard classification methods or advanced segmentation of specific thematic indices. Once extracted, this geographical information coverage was integrated within the GIS for further analysis and management of the water crisis.

The interpretation and analysis of remotely sensed data serving to identify, delineate and characterise flooded areas was based on relationships between physical parameters, such as reflectance and emittance from the features located on the ground surface: reflectance and/or emittance decreases when a water layer covers the ground or when the soil is wet; also reflectance and/or emittance increases in the red band because of the vegetation stress caused by moisture; reflectance and/or emittance changes noticeably when different temperatures, due to thick water layer, are recorded.

In the microwave region the water presence could be assessed by estimating the surface roughness, where the dielectric constant of smooth water layer surfaces is strongly correlated to the soil water content. In the case of radar images the multi-temporal techniques were considered to identify and highlight the flooded areas. This technique uses black and white radar images of the same area taken on different dates and assigns them to the red, green and blue colour channels in a false colour image. The resulting multi-temporal image reveals change in the ground surface by the presence of colour in the image; the hue of a colour indicating the date of change and the intensity of the colour, and the degree of change. The proposed technique requires the use of a reference image from the archive showing the 'normal' situation.

5. The land cover/land use mapping

The methodology for producing the land cover/land use map from medium and high-resolution images was developed by the Remote Sensing & GIS Laboratory of NMA (Stancalie et al. 2000) on the basis of the following requirements:

- the structure of this type of information must be at the same time cartographic and statistical;
- it must be suited to be produced at various scales, so to supply answers adapted to the different decision-making levels;
- Up-dating of this information must be performed fast and easily.
- The used methodology implies following the main stages listed below:
    o preliminary activities for data organising and selection;
    o computer-assisted photo-interpretation and quality control of the obtained results;
    o digitisation of the obtained maps (optional);
    o database validation at the level of the studied geographic area; and,
    o obtaining the final documents, in cartographic, statistical and tabular form.

Preliminary activities comprise collection and inventorying of the available cartographic documents and statistical data connected to the land cover: topographic, land survey, forestry, and other thematic maps at various scales.

To obtain the land cover/land use map, satellite images with a fine geometrical resolution and rich multi-spectral information have to be used. In the case of IRS and SPOT data the preparation stage consisted in merging data obtained from the panchromatic channel, which supply the geometric detail (spatial resolution of 5 m for the IRS, 10 m for the SPOT), with the multi-spectral data (LISS for IRS, XS for SPOT) containing the multi-spectral richness. In Figure 2 a flowchart for the generation of the land cover/land use maps using high resolution satellite data is presented.

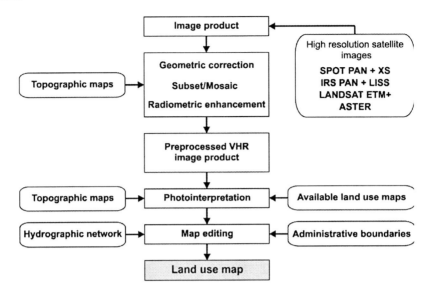

Fig. 2. Flowchart for the generation of the land cover/land use maps

For this application TERRA/ASTER data have also been used. These data proved to be suitable for detailed maps of land cover/land use, especially the visible and near infrared bands (1, 2, 3B) with 15 m resolution. The ASTER data were obtained from the Earth Observing System Data Gateway, through courtesy of Prof. Brakenridge from DFO, USA.

A series of specific image processing operations were performed with the ERDAS Imagine and ENVI software. Those operations included: geo-referencing of the data, detection of clouds and water, image improvement (through using the histogram, contrast enhancing, slicking, selective contrast, combinations between spectral bands, re-sampling operation), and statistical analyses (for the characterisation of classes, the selection of the instructing samples, and conceiving classifications). The computer-assisted photo-interpretation finalised the delineation of homogeneous areas from images, in their identification and framing within a class of interest. Discriminating and identifying different land occupation classes rely on the classical procedures of image processing and lead to a detailed management of the land cover/land use, followed by a generalising process, which includes:

- identification of each type of land occupation, function of the exogenous data and of the "true-land" data, establishing a catalogue;
- delineation of areas expected to represent a certain unit of the land; and,

CONTRIBUTION OF EO DATA TO FLOOD MANAGEMENT   299

- expanding this delineation over the ensemble of image areas, which display resembling features.

Validation of results from photo-interpretation, mapping (by checking through on-land sampling at local and regional levels) and building up the database strives to determine the reliability level and the precision obtained for the delineation of the units and their association with the classes in the catalogue.

The satellite based cartography of the land cover/land use is important because it facilitates periodical updating and comparisons, and thus contributes to characterisation of the human presence and provides information on vulnerability and serves for evaluation of the impacts of flooding.

The land cover maps are useful to classify the terrain functions of the main types of land cover, thus allowing their characterisation with respect to the degree of land imperviousness, of their absorption capacity or resilience of water infiltration into soils.

6. Method for the identification and mapping of the flooded areas

The methodology for the identification, determination and mapping of the areas affected by floods is based on a different classification procedure of optical and radar satellite images.

The advantage of using high resolution optical satellite images consists in the possibility of selecting precise spatial information for the respective area (through merging images) and localising and defining the flooded or flood-prone (through classifications). Even during the periods with abundant rainfalls, radar images can provide useful information regarding the flooded areas. The multi-temporal image analysis, combined with the land cover/land use information allows the identification of the area covered by water (including the permanent water bodies), including the flooded areas.

Figure 3 presents a flowchart for generation of the flood extent maps using satellite radar (SAR) images. Using this methodology for the identification and mapping of the flooded areas, it is possible to monitor and investigate the flood evolution during various phases. Figure 4 shows an example of the utilisation of optical data for the flood extent mapping in the Crisul Alb basin, using MODIS-TERRA satellite data of April 9, 2004. This approach is very useful especially after the crisis, for conducting a damage inventory and for recovery actions, designed to re-build destroyed or damaged facilities and to undertake adjustments of the existing infrastructure.

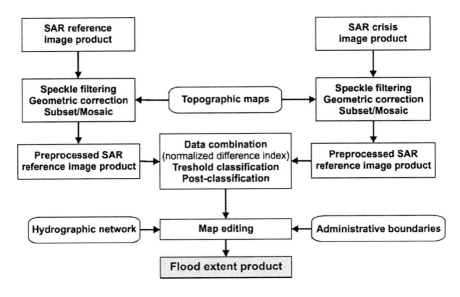

Fig. 3. Flowchart for the generation of the flood extent maps using satellite radar (SAR) images

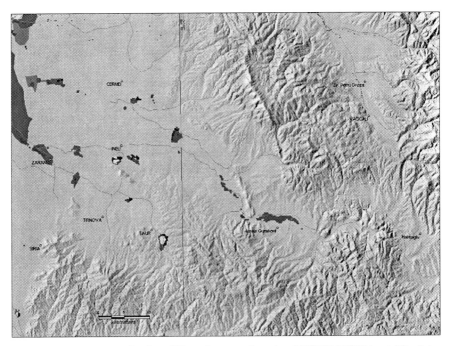

Fig. 4. Flood extent mapping in a GIS environment based on MODIS-TERRA satellite data of April 9, 2004

7. Method for the flood risk maps preparation

The assessment of the flood risk hazard requires a multidisciplinary approach coupled with the hydrological/hydraulic modelling. The contribution of geomorphology can play an exhaustive and determining role when using the GIS tools (Townsend and Walsh 1998).

The structure of the GIS was planned for use in this study, for evaluation and management of information that contributes to flooding occurrence and development, as well as for the assessment of damages inflicted by flooding. In this regard the database represented by the spatial geo-referential information ensemble (satellite images, thematic maps and series of the meteorological and hydrological parameters, and other exogenous data) is structured as a set of file-distributed quantitative and qualitative data focused on the relational structure between the info-layers. The GIS database will be connected with the hydrological database, which will allow synthetic representations of the hydrological risk, using separate or combined parameters. Figure 5 displays the procedure for integrating the hydrologic/hydraulic model outputs and GIS info-layers for preparation of flood risk maps.

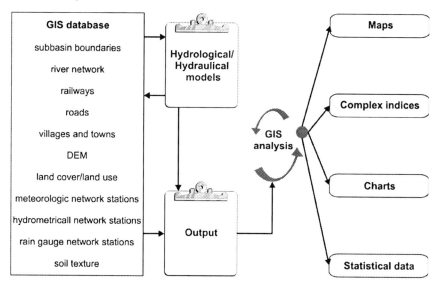

Fig. 5. Integration of the hydrologic/hydraulic model outputs and GIS info-layers for preparation of flood risk maps

The construction of the GIS for the study area of the Crisul Alb and Crisul Negru basins was based mainly on classical mapping documents, such as maps and topographic plans. Most of the thematic plans will be extracted from this classical mapping support. Due to the fact that, in most of the cases, the information on the maps is outdated, it is planned to update it using recent satellite images (e.g., the hydrographic network, land cover/land use) or by field measurements (e.g., dykes and canals networks). The GIS database contains the following info-layers: sub-basin and basin boundaries; land topography (organised in DEM); hydrographic network, dykes and canal networks; transportation network (roads, railways), municipalities, meteorological station network, rain-gauge network, hydrometric station network; and land cover/land use, updated from satellite data.

In order to produce the flood risk map for the test-area, an important step comprised construction of the DTM and its integration, together with the land cover/land use maps, in the GIS using a common cartographic reference system.

The DEM was produced by taking the following steps:

- scanning the topographic maps at 1:5,000 and 1:10,000 at 300 dpi resolution;
- geo-referencing the maps in the UTM projection;
- colour separation and raster information layer extraction as a linear image (black & white without grey tones);
- vectorisation of raster images.
- merging the maps; and,
- generating a triangular irregular network (TIN) model.

In a TIN the point density on any part of the surface is proportional to the variation in terrain. A surface is a continuous distribution of an attribute over a two-dimensional region. TIN represents the surface as contiguous non-overlapping triangular faces. The surface value is estimated in this way for any location by simple (or polynomial) interpolation of elevation in a triangle. As elevations are irregularly sampled in a TIN, it was possible to apply a variable point density to areas where the terrain changes sharply, yielding an efficient and accurate surface model.

Various morphological criteria may be extracted from a DEM, designed using topographical maps and geodetic measurement criteria, such as altitudes, slopes, exposures, transversal profiles and thalweg locations. All these parameters are useful to evaluate the local or cumulated potential of inflow into a zone of the basin. This approach is also useful to get a realistic simulation of floods, taking into account the terrain topography, the hydrological network and the water levels in different transversal profiles on the river, obtained from hydraulic modelling.

Using the GIS database for the study area that includes the DTM, the land cover / land use maps and the vector info-layers (hydrographical, dams and canals networks, the transportation and municipalities network, etc.) several simulation outputs of hydrological or hydraulic models could be superimposed in order to produce the flood risk maps.

8. Conclusions

Flood risk analysis needs to use and integrate many sources of information. This approach is more demanding in the case of a transboundary river. The integrated flood management approach is in agreement with the recommendations of the International Strategy for Disaster Reduction and the EU Best Practices on Flood Prevention, Protection and Mitigation (Balint 2004).

Although satellite sensors cannot measure the hydrological parameters directly, optical and microwave satellite data supplied by the new European and American orbital platforms, such as the EOS-AM "Terra" and EOS-PM "Aqua", DMSP, Quikscat, SPOT, ERS, RADARSAT, and Landsat7, can supply information and adequate parameters to contribute to the improvements of hydrological modelling and flood warning.

Considering the necessity to improve the means and methods of flood hazard and vulnerability assessment and mapping, the paper presents the capabilities offered by remotely sensed data and GIS techniques to manage floods and the related risk. The study area is situated in the Crisul Alb - Crisul Negru - Kőrös transboundary basin, crossing the Romanian - Hungarian border.

The specific methods, developed in the framework of the NATO SfP project "Monitoring of extreme flood events in Romania and Hungary using EO data" for deriving satellite-based applications and products for flood risk mapping (maps of land cover/land use, thematic maps of the flooded areas and the affected zones, flooding risk maps), were also presented.

The satellite-based applications will contribute to preventive consideration of the extreme flood events by planning more judiciously land-use development, by elaborating plans for food mitigation, including infrastructure construction in the flood-prone areas, and by optimisation of the flood - related spatial information distribution to end–users. At the same time the project will provide the decision-makers with updated maps of land cover/land use and the hydrological network, and with more accurate/comprehensive thematic maps at various spatial scales indicating the extent of the flooded areas and the affected zones.

References

Balint Z, Toth S (2004) Flood protection in the Tisza basin. Proc. of the NATO Advanced Research Workshop – Flood Risk Management Hazards, Vulnerability, Mitigation Measures, Ostrov u Tise, Czech Republic, pp 171-183

Brakenridge GR, Stancalie G, Ungureanu V, Diamandi A, Streng O, Barbos A, Lucaciu M, Kerenyi J, Szekeres J (2001) Monitoring of extreme flood events in Romania and Hungary using EO data. NATO SfP project plan, Bucharest, Romania

Brakenridge GR, Anderson E, Nghiem SV, Caquard S, Shabaneh TB (2003) Flood Warnings, Flood Disaster Assessments, and Flood Hazard Reduction: The Roles of Orbital Remote Sensing. Proc of the 30th Int. Symp. on Remote Sens. Environ, Honolulu, Hawaii

Samuels PG (2004) An European perspective on current challenges in the analysis of inland flood risks. Proc. of the NATO Advanced Research Workshop – Flood Risk Management Hazards, Vulnerability, Mitigation Measures, Ostrov u Tise, Czech Republic, pp 3-12

Stancalie G, Alecu C, Catana S, Simota M (2000) Estimation of flooding risk indices using the Geographic Information System and remotely sensed data. Proc. of the 20th Conference of the Danubian Countries on hydrological forecasting and hydrological bases of water management, Bratislava, Slovakia

Townsend P, Walsh SJ (1998) Modeling of floodplain inundation using an integrated GIS with radar and optical remote sensing. Geomorphology 21:295-312

ON-LINE SUPPORT SYSTEM FOR TRANSBOUNDARY FLOOD MANAGEMENT: DESIGN AND FUNCTIONALITY

V. CRACIUNESCU[1], G. STANCALIE
Romanian Meteorological Administration, 97 Soseaua Bucuresti-Ploiesti, 013686 Bucharest, Romania

Abstract. An important objective of the NATO SfP project "Monitoring of extreme flood events in Romania and Hungary using EO data" is the development of a dedicated sub-system, based on remote sensing and GIS technology, serving to improve the flood management and implementation of mitigation programs. The sub-system consists of a GIS database, represented by the spatial geo-referential information ensemble connected to the classical hydrological database, and of a hydrological forecast model. The main functions of this sub-system are: (a) acquisition, storage, analysis, management and exchange of raster and vector graphic information and the related attribute data for flood monitoring activities, (b) updating the information and restoring data, (c) development of thematic documents, and (d) generation of value-added information and products.

Keywords: flood, hazards, vulnerability, risk, NATO, spatial information, satellite data, on-line support system, GIS database, spatial information, land cover/land use, mapping, digital elevation model

1. Introduction

Flood management evolves and changes as more knowledge and technology becomes available to the environmental community. One of the most powerful tools to emerge in the hydrological field is the Geographic Information System (GIS), which allows the collection and analysis of

---

[1] To whom all correspondence should be addressed. Vasile Craciunescu, National Meteorological Administration, 97 Sos. Bucuresti-Ploiesti, sector 1, 013686, Bucharest, Romania; e-mail: vasile.craciunescu@meteo.inmh.ro

environmental data. The decision process starts with observed data that support the generation of information through modelling, such information then evolves into the knowledge through visualisation and analysis, and finally this knowledge supports hydrological decisions.

In recent years, integrated remote sensing data started to play an important role in the creation or updating of the existing GIS databases. Earth Observation (EO) images have wide applications in flood analysis, in such tasks as producing catchments maps, detecting water surface and soil moisture, detecting inundated areas, and assisting with remote flow measurement. Thus, image processing is important for developing such products and using them in flood analysis and management (Brakenridge et al. 2001).

The distribution of the graphic and cartographic products (derived using the GIS facilities on the basis of remote sensing data, maps and field surveys) to the interested authorities, media and the public is an important issue in the framework of the NATO SfP project "Monitoring of extreme flood events in Romania and Hungary using EO data". These products will contribute to preventive consideration of flooding in land development and special planning for the flood-prone areas, and for optimising the distribution of flood-related spatial information to end-users (Brakenridge et al. 2003).

The paper presents the design and the main functions of the flood monitoring on-line support system for spatial information management, as well as the preliminary results of the system implementation.

2. Study area: Crisul/Körös basin

The study area represents the Crisul Alb/Negru/Körős transboundary basin spanning across the Romanian-Hungarian border, with a total area of 26,600 km^2 (14,900 km^2 on the Romanian territory).

In Romania, the catchment (Fig. 1) comprises mountainous areas (38%), hilly areas (20%) and plains (42%). About 30% of the catchment is forested. On the Hungarian side, the catchment relief represents plains. Annual precipitation ranges from 600-800 mm/year in the plain and plateau areas to over 1200 mm/year in the mountainous areas of Romania. This precipitation distribution can be explained by the fact that humid air masses brought by fronts from the Icelandic Low frequently enter this area. The orography of the area (Apuseni Mountains) amplifies the precipitation on the western side of the mountain range. Thus, the Crisuri Rivers Basin frequently experiences large precipitation amounts in short time intervals and the frequency of such events seem to be increasing in recent years (Brakenridge et al. 2001).

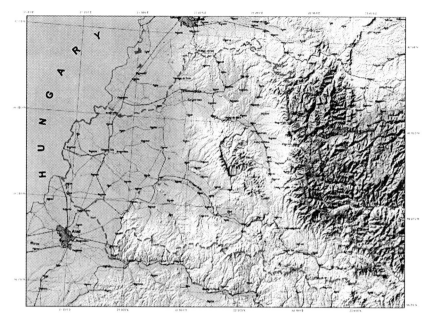

Fig. 1. Romanian part of the Crisul Alb/Negru basin

3. On-line support system for spatial information management

A flood forecasting and warning system is already active in the study area. The existing system does not include a spatial component of the phenomena (pertaining to geographic distances or patterns) both in the pre- and post-crisis phases. The purpose of the development of a dedicated sub-system based on remote sensing and GIS technology is to contribute to the regional quantitative risk assessment for monitoring and hydrological validating of risk simulations, in the Romanian-Hungarian transboundary test area. Also an important result will be the preventive consideration of flood events when determining land development and in special land use planning for the flood-prone areas (Brakenridge et al. 2003).

The main function of this sub-system will be:

- acquisition, storage, analysis and interpretation of data;
- management and exchange of raster and vector graphic information, and also of related attribute data for flood monitoring activities;
- handling and preparation for a rapid access of data;
- updating the information (temporal modification);
- data restoring, including the development of thematic documents;

- generation of value-added information (complex indices for flood prevention, risk maps); and,
- distribution of the derived products to the interested parties, media, etc.

The proposed dedicated sub-system based on remote sensing and GIS is presented in Fig. 2, with the data flows and the links between the data suppliers and end-users.

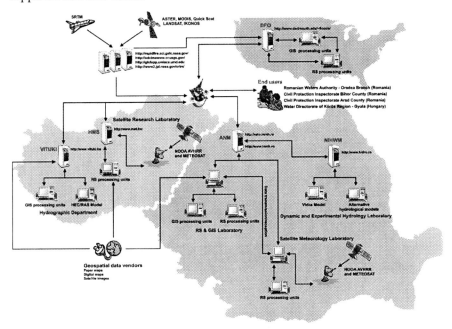

Fig. 2. Dedicated sub-system based on remote sensing and GIS technology

3.1. CONSTRUCTION OF THE GIS DATABASE

The structure of the dedicated GIS database was planned to be used for the study, evaluation and management of the information, which relates to the flooding occurrence and development, as well as for the assessment of damages inflicted by flooding. In this regard the database represented by the spatial geo-referential information ensemble (satellite images, thematic maps, series of the meteorological and hydrological parameters, and other exogenous data) is structured as a set of file-distributed quantitative and qualitative data focused on the relational structure between the info-layers. The GIS database is connected with the hydrological database, which allows synthetic representations of the hydrological risk using separate or combined parameters (Stancalie et al. 2003; Brakenridge et al. 2004).

It was decided to develop a GIS database for the whole study area of the Crisul Alb, Crisul Negru and Kőrős basins using different cartographic documents with the scale of 1:100,000. The construction of this GIS is based mainly on classical mapping documents, specifically represented by maps and topographic plans. Most of the thematic layers were extracted from this classical mapping support. Due to the fact that, in most of the cases, the information on the maps was outdated, it was required to update it on the basis of recent satellite images (e.g. the hydrographic network, land cover/land use) or by field measurements (e.g. dykes and canals network). The topographic maps at 1:100,000 scale in Gauss-Kruger projection (zone 34) contain the information needed to construct the GIS database for the whole study area.

The GIS database contains the following info-layers: (a) sub-basin and basin limits; (b) land topography (90 metres DEM); (c) hydrographic network, dyke and canal networks; (d) transportation network (roads, railways); (e) municipalities; (f) meteorological stations network, rain-gauge network, and hydrometric stations network; and, (g) land cover/land use, updated from satellite images.

The GIS info-layers related to the hydrographical network and the road and railway network in the Crisul Alb and Crisul Negru Romanian basins are presented in Figure 3.

The preparation of the info-layers that constitute the digital geographic information database or the geo-spatial information was achieved by:

- identification of the reference points;
- scanning the cartographic documents (on paper);
- integration of the geo-spatial information in the thematic info-layers; and
- association of attributes for different geographic objects (watercourses, meteorological and hydrological stations, villages and towns, roads and highways, etc.).

For the acquisition of digital geographic data, it was necessary to specify the information layers related to:

- the scale of the cartographic documents or image data;
- the type of the geographic objects, which constitute the layers (represented by layers in vector, tin or raster formats);
- the attributes which characterised them; and,
- the file format and coordinate system.

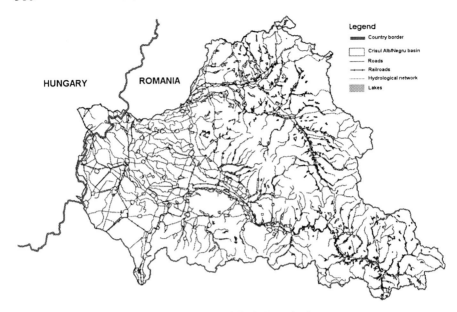

Fig. 3. GIS info-layers for the Crisul Alb and Crisul Negru basins

For the most flood vulnerable areas considered, located in the plain of the Crisul Alb/Negru/Körös basins and delineated at its Eastern boundary by the Ineu –Talpos and its northern boundary by the Crisul Repede basin (Figure 4), a more precise GIS database was constructed using 1:5,000 and 1:10,000 topographic plans.

One of the most important products obtained for this area is a precise digital elevation model (DEM). For this purpose the shape with elevation information extracted from individual map sheets have been merged and corrected, and then interpolated to obtain the DEM.

Interpolation methods produce a regularly spaced, rectangular array of Z values from irregularly spaced XYZ data. The term "irregularly spaced" means that the points follow no particular pattern over the extent of the map, so there are many "holes" where data are missing. Interpolation fills in these holes by extrapolating or interpolating Z values at those locations where no data exist (Lee and Schachter 1980; Isaaks and Srivastava 1989).

The interpolation methods tested were based on Kriging, Triangulated Irregular Network (TIN), Minimum Curvature and Natural Neighbour algorithms. The best result was obtained for the Kriging method. The digital elevation model was then used for deriving the terrain slope, aspect and curvature maps.

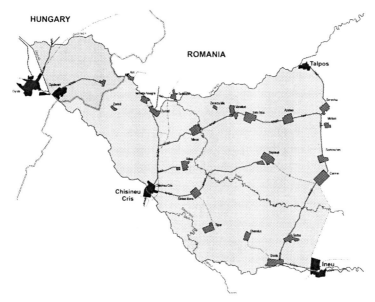

Fig. 4. Vulnerable areas in the Crisul Alb/Negru/ Körös basin

3.2. PREPARATION OF SPATIAL DATA FOR RAPID ACCESS

The project objectives include working with different types of spatial geo-data (scanned maps, satellite images, vector files, digital elevation models) in different file formats and coordinate systems, processed by the project partners in Window, Linux, and Solaris computing platforms and software environments. To make all information easily available to the participants and end-users, a detailed specification package was developed. This package ensures that every piece of information uses the same file format (ESRI shapefile for vector data; ESRI grids for the digital elevation model; ERDAS .img for maps and satellite images) and the same geographic coordinate system: UTM Zone 34/WGS84 (Figure 5).

At this point one of the most important tasks was to build a Satellite Image Database (SID), which would gather information on the available raw satellite scenes and the derived products, and make it available in a simple format. The SID was build in MySQL and is available on-line on a server, being updated as new satellite images are acquired. Each record of the database describes the characteristics of each satellite image: platform, sensor, date and time of data acquisition, duration of pass, spectral band, coordinates of the area covered, projection, calibration, size, bits/pixel, image file format, physical location (machine, directory), origin of data, type (raw/processed), type of processing applied, algorithm used, quick-look available, and cloudiness. Queries are very easy to conduct using the web interface.

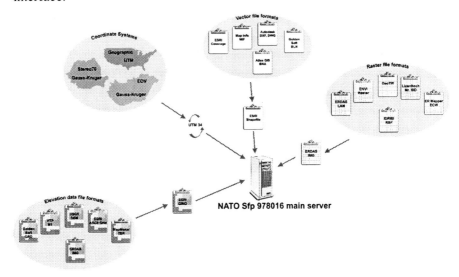

Fig. 5. Spatial data preparation for rapid access

3.3. SPATIAL DATA DISEMINATION

One of the most important functions of the dedicated sub-system is the distribution of the project results to the participants, end-users and the public. The easiest way to distribute the spatial and tabular attribute data is by setting up a FTP server where the information could be stored and accessed. From the end-user point of view, this approach has two major disadvantages:

- As the database grows the relevant information is more difficult to find;
- the data are stored in a common GIS file format, which may require special software and training for the user be able to read and analyse the data.

Another option is to distribute spatial and tabular attribute data over an Internet Web-based network, which is a powerful and effective communication method overcoming the disadvantages of the first approach and allowing all interested agencies and end-users data access without being technical experts.

Generally, viewing GIS data on the Web involves a three-tiered architecture:

- a spatial server that can efficiently communicate with a Web server and is capable of sending and receiving requests for different types of data from the Web browser environment;
- a mapping file format that can be embedded into a Web page; and,
- a Web-based application in which maps can be viewed and queried by an end-user/client via a Web browser.

Publishing the data on the Web using this approach would not change the existing data workflow, i.e., how the data are created, maintained, and used in desktop applications (Hendry 2004). This means that the map-server dynamically generates maps from the files stored in a certain folder every time a user sends a request. The Web-based application was developed using standard technologies such as HTML, XML, JavaScript, PHP, SVG, COM and supports the Open GIS Consortium (OGC) and the Open Web Services specifications (Fig. 6).

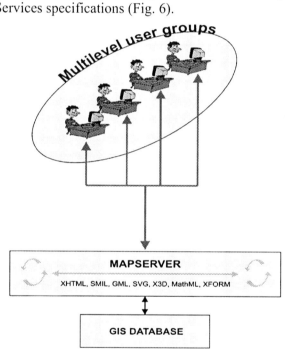

Fig. 6. Spatial data distribution over the Internet

4. Conclusions

The development of a dedicated sub-system based on remote sensing and GIS technology will improve the flood management and will aid the implementation of flood mitigation programs in the Romanian-Hungarian transboundary study area. This sub-system will allow the storage, management and exchange of raster and vector graphic information, and also of related attribute data for flood monitoring activities. This dedicated information sub-system will contribute to regional quantitative risk assessment (using flood hazard and vulnerability characteristics) for monitoring and hydrological validating of risk simulations. Another important result will be the preventive consideration of flood events when determining land development and in special land use planning in the flood-prone areas.

The GIS database for the selected area in the Crisul Alb and Crisul Negru basins will be implemented at the Crisuri Rivers Authority, Oradea and the District Inspectorates of Civil Protection in Arad and Bihor. The GIS database will be set up at the local operational hydrological service and at the District Inspectorate of Civil Protection operational units. Communications of information between the National Meteorological Agency and the Crisuri Rivers Authority, as well as between the Water Authority and the District Inspectorate of Civil Protection will use the FTP or the e-mail for a simple mail transfer protocol to upload and download data and other information.

References

Brakenridge RG, Stancalie G, Ungureanu V, Diamandi A, Streng O, Barbos A, Lucaciu M, J Kerenyi, Szekeres J (2001) Monitoring of extreme flood events in Romania and Hungary using EO data, project plan (September). Bucharest, Romania

Brakenridge RG, Stancalie G, Ungureanu V, Diamandi A, Streng O, Barbos A, Lucaciu M, J Kerenyi, Szekeres J (2003) Monitoring of extreme flood events in Romania and Hungary using EO data. Progress report (May). Hanover NH, USA

Brakenridge RG, Stancalie G, Ungureanu V, Diamandi A, Streng O, Barbos A, Lucaciu M, J Kerenyi, Szekeres J (2004) Monitoring of extreme flood events in Romania and Hungary using EO data. Progress report (May). Hanover NH, USA

Stancalie G, Diamandi A, Ungureanu V, Stanescu VA (2003) Sub-system based on remote sensing and GIS technology for flood and related effects management in the framework of the NATO SfP "TIGRU" project. In: Proc. First Annual Session of the NIHWM, Bucharest, Sep. 22-25 (CD-Rom)

Lee DT, Schachter BJ (1980) Two algorithms for constructing a Delaunay Triangulation. International Journal of Computer and Information Sciences 9(3):219-242

Isaaks EH, Srivastava RM (1989) An introduction to applied geostatistics. Oxford University Press, New York

Stancalie G, Alecu C, Craciunescu V, Diamandi A, Oancea S, Brakenridge GR (2004) Contribution of Earth Observation data to flood risk mapping in the framework of the NATO SfP "TIGRU" project. In: Proc. BALWOIS, Int. Conference on Water Observation and Information System for Decision Support, Ohrid, FY Republic of Macedonia, May 25-29 (CD-Rom)

Hendry F (2004) Best practices for web mapping design. Proc. of The second MapServer Users Meeting, Ottawa, Canada June 9-11(CD-Rom)

TERRITORIAL FLOOD DEFENSE: A ROMANIAN PERSPECTIVE

MARCEL LUCACIU[1]
Arad County Emergency Situation Inspectorate, Arad, Romania

Abstract. Floods, especially those occurring on large rivers, occur over areas which are greater than those controlled by a single administrative territorial unit. The flood hydraulic regime is influenced by the extent and quality of flood defence structures. All the adopted measures influence the forming and evolution of the flood wave during the flood situation. This requires that the flood protection structural measures, adopted at the basin level, have to be integrated and correlated with all the measures adopted by the local authorities through the flood defence plans. These plans better reflect the level of risk exposure and the real possibilities of the communities to deal with flood situations. Such requirements become even more obvious in border regions, where, without an efficient and fast exchange of information, one of the parties can be badly affected by the floods occurring on the other party's territory.

Keywords: vulnerability and risk reduction, structural and non-structural flood protection measures, national system for emergency situations, emergency situation plans, flood defence plans, monitoring, transboundary cooperation

1. Introduction

As anywhere else, the flood defence in Romania is a matter of national interest. Floods are considered natural disasters. A dangerous event, triggered naturally, becomes a disaster when its effects on human life, properties and environment, with major consequences.

---

[1] To whom all correspondence should be addressed. Marcel Lucaciu, Arad County Emergency Situation Inspectorate, Arad, Romania; e-mail: m_lucaciu@hotmail.com

One can not precisely establish levels or thresholds, over which a dangerous event becomes a disaster, because this consequence depends on socio-economic context in which it takes place. But there are some keywords, which can allow us to define a disaster:

- The event has to produce adverse effects in terms of population, property, or environment;
- The magnitude and the scale of affected areas or communities are significant; and,
- The local resources for dealing with event are overwhelmed.

Floods represent a major risk for Romania, especially for the western part (Chiriac and Manoiu 1980). A particular issue for floods occurring in this part of the country is the transboundary effect; the fact that they can affect the communities in Hungary. This determines the need to adopt common defence measures in both countries, so that the measures applied in Romania do not increase the magnitude and the level of danger and losses in Hungary. This can be done, on the one hand, through common hydrotechnical works (which will diminish the effects and allow each party to adopt its own efficient structural defence measures), and, on the other hand, to organise, at the regional level, common defence forces and activities, in order to synchronise efforts and make them much more efficient.

This paper will present the territorial flood defence from the Romanian perspective in the hydrological basins of the three main rivers, the Mures, the Crisul Alb and the Crisul Negru, which cross over the Arad county Territory. It will also identify some ideas for a better integration of the Romanian and Hungarian efforts from this perspective.

In order to better understand the Arad county territorial flood defence, one needs to first describe the flood risk area (represented by the three main rivers and their tributaries) and its vulnerability. Next, the structural and non-structural flood defence measures will be described, and finally, the functional and administrative organisations for flood defence will be presented with a special emphasis on transboundary cooperation.

2. Study area

Arad County is located in the Western part of Romania, at the border with Hungary and Serbia and Montenegro. The total population is about 500,000 inhabitants living in 78 settlements, including a larger municipality, 9 towns and 68 villages. The most important urban centre is Arad with over 180,000 inhabitants. Geographically the territory is spread in three equal parts among plains, hills and mountains. Forty percent of the total county territory is covered with forests. There are four important national roads with a total length of over 500 km, four border crossing points (three for vehicles and one for trains), and a main railroad which links Romania and Hungary. The county is economically developed. The main economical domains are industry representing 37%, construction 4.1%, transport and communications 7.3%, services 15.7% and agriculture, representing 20.4%. The most important industrial branches are railway industry, light industry (textile, leather, shoes), civil construction, and food industry. The most important natural resources include metal ore, construction materials and oil. The county is crossed over by three main rivers, the Mures, the Crisul Alb and the Crisul Negru, with the last one crossing only over 15 km.

2.1. HYDROMETEOROLOGICAL CHARACTERISTICS OF THE AREA

2.1.1. *Mures River*

The Mures River is one of the most important Romanian watercourses. The total catchment area is about 27,920 km^2 and the total length is 754 km on the Romanian territory, and 43 km on the Hungarian side up to its confluence with the Tisa River. The flood defence line is represented by over 190 km of dykes, most of them on Arad County's territory. The river has 11 main tributaries in Arad County.

The flood waves on this river usually have a medium duration, between 7 and 30 days, and are influenced by all the tributaries in the basin, the soil moisture, and unhindered water movement. The average flow is 169 m^3/s and the maximum flow of 2,088 m^3/s was observed at Arad in 1970. During floods, the Tisa River may affect river flows in the entire region.

Flood control structures have been constructed since the 19th century and, year by year, maintained, modernised and raised up to the current level. This can become a serious problem during a long period of flooding, when dykes can easily fail along the interface of layers from different periods of construction. The areas protected by dykes in the Arad and Timis counties measure more than 155 ha. The high level of river meandering also determines a large main bed, which made the construction of dykes even more complicated.

After the 1970 flood, new protection dykes have been constructed, especially in the flood-prone areas, with a total length of 90 km. In some areas the weak dykes made the authorities to build a second line of flood defence. The county location at the river's low point provides sufficient time for local authorities to implement flood defence measures.

Arad is the most important municipality in the area and requires the main flood defence effort. In the case of major floods, the city can be badly affected. The flood line of defence consists in 46 km of earthen and concrete dykes; along these dykes, there are 18 critical points that have to be monitored during floods.

The largest floods (in the large flood severity class) occurred in 1970, 1975, 1980, 1985, and 1998; with recurrence periods between 5-15 years. In the border area there is a good cooperation between water management organisations in Romania and Hungary.

The main causes of these floods include heavy rainfall in the entire river catchment area, especially upstream from the Arad County; sudden snowmelt in the mountainous area, as a consequence of a rapid increase in air temperatures; and, short torrential rains, which cause flash floods on the main streams (Brasoveanu 1968).

The floods occurred particularly during the spring and autumn periods, but, statistically speaking, they can occur anytime during the year.

There are some flow control works on the Mures River, by small and medium size reservoirs, which allow flood control, especially at the confluence of the Mures with its main tributaries. Three temporary storage reservoirs on the Arad County territory (total storage of 2.3 million m^3) are also used to control floods.

On the Arad River's flood plains there are some settlements at risk in the unprotected area, and 16 settlements are protected by dykes. On the Mures River there is a very important natural conservation area, The Lunca Muresului reservation, which was created in collaboration with the Hungarian counterpart and EU support.

2.1.2. *The Crisul Alb River*

With a total length of 234 km (163 km in the Arad County and 71 km in Hungary) and a catchment area of 4,486 km^2, the Crisul Alb River is the second most important watercourse in the county. It can cause bigger problems than the Mures, due to the short distance between the catchment area and the point of overflow, and also because of the short time required for flood formation. As a consequence, the river has a higher speed of runoff, which makes it more dangerous.

Of the total river length, 66 km are protected by dykes. The average water flow is 27.0 m^3/s and the maximum of 820 m^3/s was observed at Chisineu Cris in 2000. The river has 71 main tributaries that contribute the flow of more than 100 m^3/s during floods.

The flood waves on this river usually have a medium duration, between 2 and 8 days, and are influenced by all the tributaries in the basin, the soil moisture level and the unhindered flow of water. During floods, the water level in the Tisa River modifies the river flows and affects their regime. A large number of bridges, in Romania and Hungary, also influence the river hydraulic regime. A special channel was created to compensate this effect. This channel itself represents a risk factor for the neighbouring communities.

On the Cris River, there are three different hydraulic zones (Oprea et al. 1971):

- The upper basin, where efficient control of tributary flows is provided by dam reservoirs;
- An intermediate region, non-protected, where the river is free to spread flood flows, with a powerful attenuation of the flood wave; and,
- The dyke area, in the lower basin, where a lot of temporary storage reservoirs (polders) were established control flows in emergency situations. The dykes in this area increase the speed of flow.

There are some dams on the main tributaries providing flood and flow control measures. The Taut reservoir, storing over 33 millions m^3 of water, reduces the contributions to the main riverbed during floods.

In Hungary, the Crisul Alb River merges with the Crisul Negru River, which increases the flood risk in this area. A large emergency reservoir (polder) located between the rivers (in Hungary), in the case of use, strongly modifies the river flows. All the Romanian flood protection works are built in cooperation with those existing in Hungary, so one can talk about an integrated structural flood protection system.

The flood protection works have been constructed since the 19th century (1864) and, year by year, were maintained, modernised and increased in height. The town of Chisineu Cris (over 10,000 inhabitants) requires a main flood defence effort. The town was badly flooded in 1966, when a half of the town was inundated. A dyke line of defence, with a total length of 12 km, protects the town. In some sections, the dykes are reinforced with concrete plates.

The most important floods on this river occurred in 1966, 1970, 1980, 1985, 1995, and 2000, with the recurrence periods (medium flood severity class) of about 5-10 years.

On the river flood plains, there are 10 settlements at risk in the unprotected area and 9 settlements in the protected areas with dykes.

The last most important flood occurred in April 2000. This flood repeated, with some small exceptions, the situation which occurred in 1970 (the dyke failed only 11 km from the breach in 1970). During this flood, damages were afflicted on railways, roads, over 20,000 ha of agricultural land, and the Adea village.

2.1.3. *The Crisul Negru River*

The Crisul Negru crosses the county territory over a short distance of 12 km, and represents the third main flood-prone river in the county. The river has its confluence with the Crisul Alb in Hungary, and its hydraulic flow regime during floods is highly affected by the backwater effects of the Tisa River. Specific to this river are very high dykes in populated areas, reaching more than 2 m above the houses' roofs. The village of Zerind (over 2,000 inhabitants) requires the main defence effort; the bridge that links Arad with the Bihor County is a supplementary risk factor.

3. Flood defence measures

The flood protection measures are planned and implemented on the basis of five fundamental preventive protection principles for flood defence (Stanescu and Drobot 2002): (a) the water is a part of the whole basin; (b) the water has to be kept in the main bed; (c) the river must be allowed to overflow (onto flood plains); (d) there is always a flood risk; and, (e) the action has to be focused and integrated. Based on these principles, there are two main types of flood defence measures also adopted also in the Arad County and described below.

3.1. STRUCTURAL FLOOD DEFENSE MEASURES

The structural flood defence measures represent that kind of measures which affect the hydraulic and hydrologic regime of watercourses. Because they are very expensive, they involve large investments, and cause major changes in the river biodiversity. For this reasons this kind of measures have to be planned and applied at the basin level, in order to reduce their secondary effects. The exclusive use of structural flood defence measures can not solve all the flood protection problems. There will be always a risk (residual risk) due to their limitations (safety factor, overflow probability), uncertainty in their operation and the last but not the least, because their construction also increases not only the safety level but also the risk level. Dams, dykes, embankments, channels and other protective structures are designed to provide protection against some specific level of flooding. The "level of protection" is selected on the basis of costs, community's desire of safety, potential damages, environmental impacts, and other factors. Engineers can design and construct dykes, dams and other structures providing a very high level of protection but communities tend to choose lower levels of protection because of the initial costs rather than the overall costs and benefits. The exclusive use of structural protective measures is not efficient and therefore not recommended. Generally speaking, it is necessary to choose a basic level of protection (provided by structural flood defence measures) in combination with much less expensive non-structural methods.

3.1.1. *Measures for flood flow wave reduction*

Permanent reservoirs reduce flood severity by shaving off the flood wave crest and controlling the river flow. For the three main rivers in the study area, there are four such structures: three small reservoirs on the Mures River and one on the Crisul Alb River. Only one of them, the Taut Lake, has permanent water storage, being affected by the sediment depositions. The use of these reservoirs is regulated.

Temporary storage (emergency) reservoirs (also called polders) reduce water levels during high floods, when they allow to transfer a large quantity of water from the main river channel to a lateral reservoir (polder). In the study area, there the following temporary storage reservoirs: the Mures River - none, the Crisul Alb River - 3 (24.5 million m^3 storage capacity) and the Crisul Negru River - 4 (32.7 million m^3 storage capacity) (Selariu et al. 1993).

3.1.2. *Structural measures based on soil conservation and rational utilisation*

These measures are the most efficient, because they act at the point where runoff is being generated, and work on the principle "keep the water where it falls". They are not developed within the Arad county, where middle and low sections of rivers exist. These activities are mainly developed on tributaries and consists in the following: slope terraces, forestation, river bed rehabilitation, stormwater management in urban areas, flow diversion channels and others.

3.1.3. *Inundation duration reduction techniques*

Drainage of inundated areas is provided by a very important specialised state agency playing an important role in flood protection and flood mitigation measures. This measure becomes active at the beginning of dangerous flood events and continues long after the flood ends, especially when some dyke failure occurred. In the Arad county there are 6,796 km of drainage channels and 61 pumping stations.

3.1.4. *Maximum water level reduction techniques*

Cleaning the riverbed - this activity is achieved through special programs of the water administration bodies together with local public administration. In this respect, the biggest problem encountered is the continuing reduction of the water administration staff, whilst the areas which they have to service are getting bigger and bigger. Also, these activities have to be fulfilled with a fixed small budget.

Riverbed controls are executed very rarely because of scarce founds available; usually such activities are executed after large floods.

3.1.5. *Measures protecting people and properties against floods*

The dykes and embankments represent the main structural flood protection measure. The dykes provide generally a good protection and create the false idea of a complete safety. Nevertheless, in the case of major dyke failures, the property damages and even the loss of human lives can be more severe than without dykes. The closer is the dyke located to the river channel, the worse is the harmful effect in the case of dyke failure.

One should also bear in mind the ecological effects of dykes, caused by the discontinuity between certain the river channel and flood plains. Another issue concerning the dykes is the associated flood insurance level. The dykes on the Arad rivers are constructed for an insurance level of 2% for settlements (for an extreme 50-year flood).

3.2. NON-STRUCTURAL FLOOD DEFENCE MEASURES

Due to the high costs of constructing and maintaining the flood dykes and other similar works, the non-structural flood defence measures represent efficient complementary defence solutions, which are in most of cases much cheaper. Thus, the non-structural measures are an important component of flood protection and belong to the activities supporting sustainable development (Stanescu and Drobot 2002).

The main non-structural flood defence measures adopted in the Arad county as a part of the territorial flood defence plan are discussed below.

3.2.1. *Floods peak prediction, hydrological alerts, flood warnings and forecasts*

Flood forecasts and warnings are the main task of the water management organisations at the national and basin levels. The distribution and flow of such information is organised on three levels:

- Level I: From the national or basin water management agencies to the specialised water agencies at the county level;
- Level II: from the specialised water agencies at the county level to the agencies responsible for the management of emergency situations;

- Level III: from the agencies responsible for the management of emergency situations to the local administration, which is responsible for activation of the flood defence plans.

On the three major rivers in the county, the situation from this perspective is managed as follows: both water organisations in charge of control of river sectors in the county are supported in their tasks by an automatic hydrological measurement system (automatic data reading sensors and transmitters). The system consists of 10 automatic water level gauges and their communication means. A dedicated computer program allows data interpretation and distribution to all users. This system has been operational since 2004 and was established through the DESWAT program with EU support; the old manual data collection system remains in use as a backup. To analyse and interpret the flood wave data and current hydrological and meteorological information the water management authorities are using special computer programs.

The County Operational Centre (as a part of the county emergency situation inspectorate) disseminates hydrometeorological data to the local administration. For this purpose, the inspectorate uses its own and rented communication nets and means, including: telephone, fax, e-mail, UHF radio communications, specialised network of radio receivers, and specialised computers. These means provide sufficient redundancy.

3.2.2. *Territorial management planning system*

This system comprises the following:

- Guidelines and regulations for decision-making and planning process, co-ordinated by water authorities and the county emergency situation inspectorate;
- Establishing the risk areas and the main-stream management provided by the water authorities;
- Risk mapping. This activity is organised, conducted and co-ordinated by the county council that distributes the data to the municipal governments, in order to establish the general urban plans; and,
- The prohibition to build in flood plains is a measure enforced by the specialised services of the local government, the state water authorities and the state construction inspectorate.

3.2.3. *Economic instruments serving to promote flood prevention measures*

The main economic instrument promoting flood prevention measures, in this respect, could be the national flood insurance programme. Unfortunately, even though the Romanian market is full of insurance companies, because of the lack of state involvement in this area, this very powerful tool is non-operational at this moment. A World Bank programme for implementing such a system is being introduced in Romania.

3.2.4. *Institutional reform*

The reform consists of three activities listed below.

- Building a legal frame for the functioning and co-ordination of the responsible institutions for making the strategies and the operational decisions at the basin and national levels (those concerning urban planning and development, as well as those directly managing waters). Such a specific action (the law for The National Emergency Management Situation System) was approved in 2004 and is being now implemented.
- Regulating the planning process, which comprises: establishing the intervention procedures for different scenarios and different types of events of various magnitudes, as well as, establishing the principles of civil-military co-operation at the institutional level. A main principle in making these programmes efficient is the separation of the actions reducing material losses from those reducing losses of human lives. For the latter objective, the programmes concerning good functioning of the automatic flood warnings (based on hydrological information) and good communication networks are the key elements.
- Public awareness and preparedness programmes and campaigns publicising flood defence are planned and carried out by state authorities; every two years the county inspectorate for emergency situations plans and executes warning-alert exercises for the communities in the areas with high risk of flooding.

4. Strategies for flood defence preparedness

The strategy adopted to provide flood protection, by implementing structural and non-structural measures, is implemented by the water authorities as well as the local public administration in three different dimensions:

- Communities and environment vulnerability reduction measures (preparedness phase) through: (a) restrictions on land use in flood-prone areas and implementation of individual and multiple precautionary measures (restrictive legislation, land use mapping, flood risk maps, providing intervention assets, etc.); (b) setting up warning and information systems and also planning flood defence for each

community, and conducting pro-active programs for public information and training.
- Measures for reduction of the magnitude and frequency of the flood wave (mitigation and preventive phase). This phase is based especially on structural protective methods (e.g. building dykes according to the national standards and as per negotiations with local population).
- Rapid response measures when floods occur, recovery and rehabilitation measures after the flood (response, recovery and rehabilitation phase) (UN 1993).

The responsible ministries (The Ministry of Water, Forests and Environment and The Ministry of Administration and Interior) regulate the planning activity for territorial flood defence, consisting in:
- Preventive and preparedness measures for intervention before floods occur;
- Urgent operational measures when flood starts (warning and alerting the local authorities and population, intervention, search & rescue operation, evacuation, care for evacuees, dam and dykes control);
- Post-flood intervention measures for recovery and rehabilitation.

In order to provide a timely reaction of all the systems, when flood starts, according to the Romanian legislation, there are three different situations, described by their level of threats to the life and property's safety, defined as (Stanescu and Drobot 2002):
- Attention situation has the significance of a special situation and does not necessarily imply a danger.
- Alert situation describes the evolution of the phenomena towards a certain danger (the rise of water level, flow);
- Danger situation (emergency) is declared when the threat/danger becomes imminent and some exceptional measures are needed.

Protection and intervention measures are planned according to the three flood defence alert phases. The decision to declare one of these alert phases is made by the water management authority or the county inspectorate for emergency situation. These alert phases are communicated immediately to local authorities that are in charge of their implementation.

The protective and intervention measures are planned on the basis of the river levels (stages) of danger. These levels of danger can be characterised as:
- Regional warning levels are established according to the information from raingauges the flow gauging stations located in river headwaters;

these assets provide measurements of rainfalls, water levels and flows, on the basis of which the local authorities can take some decisions in advance.
- Local warning levels are established with respect to the objectives of flood protection or the communities; their format represents water depths and levels;
- In the case of a flood situation, these warning levels are:
- For the areas protected by dykes: (a) The level of the 1st Phase of Defence, when the level of water reaches one third of the dyke height; (b) The level of the 2nd Phase of Defence, when the level of water reaches the mid point between the levels of the 1st and 3rd phases, and (c) The level of the 3rd Phase of Defence when the water level is within 0.5-1.5m of the maximum level for which the dyke was built.
- For the unprotected areas: (a) The Alert Level when the danger of flooding is possible in a short period of time, (b) The Flood Level represents the level when the first object to be protected under the protection plan is flooded, and (c) The Danger Level represents the level when special measures have to be taken: people evacuation; traffic restrictions for roads, bridges, and railways; and, special measures for dam reservoir operation.

5. Bodies and institutions involved in flood defence activities

Among the many state and local authority agencies and bodies that are involved in flood management activities, the most important are the state water management authority and the emergency situation management inspectorate.

5.1. THE WATER MANAGEMENT AUTHORITIES

State water management authorities are organised according to the hydrological basins. Some of their main tasks are related to:
- Regulating specific activities in their area of interest;
- Establishing and developing a flood defence strategy;
- Controlling the use of water resources;
- Establishing and planning the structural flood defence works;
- Coordinating the implementation of non-structural flood defence measures;
- Setting up a hydrometeorological information system; and,

- Conducting simulation exercises;

The main tasks of water management agencies in territorial flood defence are:

- Leading the flood defence Technical Support Group, which is a special unit of the County Commission for Emergency Situations;
- integrating the means and assets used in flood intervention activities;
- coordinating the flood defence planning process at the county level;
- establishing the general flood defence strategy for county territory;
- establishing the thresholds and levels of flood defence situations;
- establishing the areas, timing and the categories of people, livestock and mobile properties to be evacuated;
- establishing the flood critical points and the areas at risk for the county territory;
- providing technical assistance in flood defence planning activities to local authorities;
- providing technical assistance for setting up an checking the warning and information systems for flood alerts;
- monitoring the flood wave evolution, preparing and disseminating flood forecasts and other hydrological information; and,
- liaising with Hungarian counterparts.

5.2. STATE BODIES AND LOCAL AUTHORITIES

Starting in 2005, Romania is implementing the new National System for Emergency Situation Management. This system is based on a set of definitions that provide a uniform interpretation and application of the laws. The system consists in committees, professional bodies and volunteer structures being organised on three levels: national, county (regional) and local levels.

5.2.1. *The national level*

The national level is represented by the National Committee for Emergency Situations (NCES), which is a joint management body of several ministries. The Committee analyses and establishes the general strategy and concepts for planning and implementing the flood protection measures, integrating the efforts of several ministries, establishing and providing support, and declaring the state of alert or emergency. A specialised element at the national level is the General Inspectorate for Emergency Situations (GIES), which is a new structure created after the unification of the civil protection and fire department structures. Together with the National Operational Centre (NOC), a permanent service centre, GIES provides the monitoring, warning and information to the central and local authorities and the population in case of occurrence or the possibility of occurrence of emergency situations. NOC also provides the Permanent Technical Secretariat for NCES being in charge of planning and holding its meetings and providing logistical support for such activities.

5.2.2. *The Ministry Committees for Emergency Situations*

These committees are organised at each ministry that has to provide emergency situation management for the types of risk specific for their activity (determined by the means of specific authority and competency). Staff of other ministries can also participate, allowing better activity coordination in the case of complex emergency situations. Specialised units of these committees are the Operation Centres (with permanent activities) that ensure the monitoring and operative leadership for the activities under their competency.

5.2.3. *Emergency Support Functions (ESF)*

In the case of an emergency situation, each system component, at the national and county levels, according to an organisational structure, is responsible for some duties, related to their competency, in order to plan and execute intervention actions. According to the organisational structure, each state institution or agency can fulfil basic functions in support of the responsible agency.

5.3. THE COUNTY LEVEL STRUCTURES

At the county level, there are two bodies responsible for managing emergencies.

5.3.1. *The County Committee for Emergency Situations (CCES)*

CCES is a collegial body comprising the President (the prefect of the county), Vice-president (the president of the county council), Members, who represent the main state agencies, and Consultants, who are experts in various fields.

The CCES conducts the following main tasks:
- Informs NCES about potentially dangerous situations, their evolution, and the imminence of their threat;

- Assesses the emergency situations that have occurred and establishes actions and measures for their management;
- Declares (with the approval of the Minister of Administration and Interior) the "State of Alert" at the county level or only in the localities affected;
- Analyses and approves the county plan for providing the human, material and financial resources necessary to cope with the emergency situations.

5.3.2. *The County Inspectorate for Emergency Situations (CIES)*

CIES is a specialised body of the County Committee for Emergency Situations. Its Operational Centre (OC) provides permanent monitoring of risk factors, and the occurrence and evolution of emergency situations. The OC is permanently connected to the other operative centres organised by the various ministries in the area.

CIES conducts the following main tasks:

- Organises specific prevention activities mitigating emergency effects;
- Takes part in activities related to risks identification, recording and assessment, and develops a plan for dealing with risks on their territory, within the area of their responsibility (approved by the prefect);
- Coordinates and controls all prevention and intervention activities in the case of emergency situation;
- Monitors (through the operation centre) the appearance and evolution of emergency situations and keeps LCES informed;
- Plans and trains subordinated specialised intervention units;
- Oversees the CLES activities with respect to following the instructions and regulations for prevention and intervention in the case of emergencies;
- Sends all the necessary information (alerts, warnings, forecasts, orders, decisions, and other useful information) to the LCES;
- Maintains communication lines and provides operational communications with LCES;
- Maintains permanent liaison with operational centres from other operational state services and LCES, and neighbouring CCES;
- Centralises the requests for resources from the LCES and provides lines of logistical support to them; and,
- Creates and maintains the databases related to the risk and emergency situations within its area of responsibility;

In the case of emergency, the overall on-site control and coordination of all involved forces in intervention belongs to the incident commander, who is a well-trained person and expert in that specific field intervention, appointed by CCES (or NCES in severe flood situations). An advanced Command Post and operational staff can be provided to support the incident commander.

5.4. LOCAL LEVEL STRUCTURES

5.4.1. *Local Committees for Emergency Situations (LCES)*

LCES is a collegial leading organ created at the municipality or village level and comprising the mayor as a chairman, mayor's deputy as a vice-president, and members and experts. During a flood situation the LCES activities are coordinated by the Operation Centre, which is activated by the mayors or County Operation Centre.

LCES undertakes the following main tasks:

- Keeps CCES informed about potentially dangerous situations, their evolution, and the imminence of their threat to the community;
- Assesses the emergency situations which have occurred on its territory and establishes actions and measures for their management (a local flood defence plan);
- Declares (with the prefect's approval) the "State of Alert" at the community level;
- On the basis of local risks, analyses and approves the local plan for providing the human, material and financial resources necessary to cope with the emergency situations;
- Keeps the CCES and local council informed about its activities;
- Organises the emergency volunteer services at the local level, provides their training and logistic support;
- Sends requests to the CCES for supplementary resources when local resources are exhausted;
- Creates, maintains and uses the databases for emergency situations;
- Sends orders, dispositions and other information to the intervention units;
- Organises and coordinates the public information during the flood period;
- Integrates the intervention efforts of all intervention units (state and local intervention units, NGOs, volunteers, etc.);

- Provides medical care, mass evacuation, and alerts and warns the affected population;
- Coordinates the post-flood assessment and clean-up activities; and,
- Coordinates the international relief and organises the distribution of the aid.

6. The flood defence planning process

The flood defence planning process is an activity that is developed at all levels. The flood defence plans are integrated from bottom up. According to the National Water Management Agency regulations, risk maps, historical data, and analyses made by the specialised water authorities, the LCESs are evaluating the vulnerabilities of individual communities. The results of such evaluations are presented in as the territorial flood risk. Using this study and with technical advice from the county inspectorate for emergency situations, the LCES prepare the plans for flood defence.

The main elements of this plan include risk and vulnerability description; tasks and responsibilities of each LCES member; planned defence actions; warning levels and signals; actions to be executed at critical points; the use of forces and assets; mass care activities; medical care activities; evacuation plans; control, command, coordination and communications; logistic support; and public information.

7. The liaison and cooperation with Hungary in flood defence

The cooperation between Romania and Hungary for severe and extreme flood consequences mitigation has been ongoing for a very long time. It deals with common defence works, information systems, and intervention activities. Some of these common activities are relatively new, but others are a part of common history. Currently, one can divide the common flood defence activities into: (a) Activities specific to water management institutions, and (b) Activities specific to emergency management institutions.

7.1. ACTIVITIES SPECIFIC TO WATER MANAGEMENT INSTITUTIONS

The Joint Romania-Hungary Hydrotechnical Commission Regulation has been in effect since 1956. This special commission holds annual sessions. According to its official document the parties are involved in the following common activities:

- Common dykes and other flood defence structures: checking their functionality, construction over or under the dykes, and the points of measurements, and organising expert visits during floods.
- Information exchange, concerning water levels during floods, ice cover, the measurements taken by both parties, and the sudden rise of water levels due to reservoir releases.

7.2. ACTIVITIES SPECIFIC TO EMERGENCY MANAGEMENT INSTITUTIONS

Starting in 2003, the Treaty for mutual support in the case of disaster has been signed by the Romanian and Hungarian Governments, through their Ministries of Interior. According to this treaty the specialised bodies from both countries agreed to: Mutual support in the case of disaster, Common preparedness for intervention activities, Common plans for disasters with transboundary effects, Exchange of emergency information in the case of disaster; Reducing the border formalities for the intervention teams, and Direct links between, and joint work of, specialised agencies (county emergency situation inspectorates) in the border areas.

From the Arad County's perspective this results in periodical meetings and exchange of information, set-up of the Disaster Working Group (based on political decisions) from the DKMT Euro-region with specialists from five counties from Romania, Hungary and Serbia and Montenegro, common training sessions and exercises, and common databases.

Currently, both parties are trying to promote common development programs for disaster situations (EU financed), especially for common warning and communication networks between the Arad and Bekes counties, and the Bihor and Hajdu Bihar counties. Such activities will become more and more important for both countries, the entire region, and the counties involved.

8. Conclusions

Flood defence in Romania is certainly one of the best organised disaster response activities, having the oldest tradition. The flood defence has the following basic characteristics:

- The floods events can be relatively well and accurately predicted (concerning the time and place);
- Nevertheless, flood disasters causing damages occur due to some local reasons;

- Flood occurrences can be referenced to specific areas (sections of the protection system);
- Preparation for flood control can and should be made through concrete steps;
- Floods under normal situations are not disasters but natural phenomena; however, disaster can develop at any time (from a small event into a disaster requiring response);
- It is typical for flood defence that exceptional arrangements are only required in a given water system or a well defined region;
- In the border regions there has to be a joint effort and cooperation in disaster response and flood mitigation measures; and,
- Currently, there are sufficient provisions and conditions for a common territorial flood defence in the Romania-Hungary border areas.

In exchange for rich soil, irrigated land and convenient transportation in flood plains, throughout the history, people have had to deal with occasional flooding. Engineers have laboured thousands of years to lessen the risk, but their success in managing natural events have been mixed - often resulting in as much failure as success. Everything done in this field is intended to reduce the future consequences of flooding to the public, their property, jobs, and the local environment. Instead of investing mainly in traditional solutions, the state has to help local authorities and land owners to help themselves by offering grants for partial funding of protection of properties and minimising the flood risk. The basic idea is that managing flood risk means intensive research and development, complex projects, specific training and much, much more. From this perspective, the borders are becoming just a very small inconvenience.

References

Brasoveanu G (1968) Technical guidelines for flood defence. Bucharest
Chiriac V, Manoiu A (1980) Flood prevention and control. Bucharest
Oprea CV, Ceausu N, Dragan I (1971) Floods from 1970 and their effects on the agriculture of the Western part of Romania. Timisoara, Romania
Selariu M, Podani M (1993) Flood defence. Bucharest, ISBN 973-31-0470-1
Stanescu VA, Drobot R (2002) Non-structural flood defence management measures. Bucharest, ISBN 973-8176-16-6
UN (1993) Introduction to hazard disaster management program (first edition). UN, Geneva, Switzerland

INDEX

ADCP, 13, 19-22
ASTER, 45, 48, 54-55, 210, 216, 292-294, 298
bootstrapping, 219
Crisul Rivers (Romania), 75-78, 121, 127, 133-134, 137, 205-207, 213, 216, 217, 277-285, 289, 305-314, 317, 320
dams, 161, 279-285, 319
Danube River (Hungary), 14-17, 21, 58, 59, 61, 62
data - loggers, 30
 - satellite (Earth Observations), 11, 45-55, 208-209, 210, 211, 287-305
 - transfer, 38, 75, 307
decision making – multi-criterion, 163, 196, 204, 220, 226
decision support systems, 198
 on-line support system, 305-314
digital elevation model (DEM), 102, 139, 212, 213, 305, 310
disaster response, 315-333
dykes - failure, 133-141, 247-258
 - overtopping, 207, 238, 242, 247, 253
finite volume method, 143, 144
flood(s) - autumn, 233
 - control, 277-285
 - defence, 199, 237, 325
 - defence plans, 315-332
 - defence transboundary cooperation, 315-332
 - determining inundation, 49, 50, 55, 163, 165, 210
 - detention basins, 239, 265-273, 264
 - emergency preparedness, 159, 165, 169, 265-275, 323, 326
 - emergency reservoirs 207, 259-264, 265-275, 291, 319
 - emergency situation, 169, 265-275, 316, 319, 326-330
 - flash floods, 81, 143, 144, 318, 319

flood(s) - forecasting system, 57, 58, 61, 77, 80, 198, 270-271, 295
 - forecasts in real-time, 196, 200
 - forecasts – long term, 111, 115, 116
 - hazard, 196
 - historical, 156-169, 247-258, 281-285, 291, 317-319
 - hydrology, 158, 185, 213
 - management, 164, 167, 193, 195, 196, 207, 212, 220, 226
 - mapping, 8, 10
 - mitigation measures, 194, 207, 237, 321,
 - peak attenuation by storage, 277, 282-285
 - protection measures – non-structural, 168, 200, 322-324
 - protection measures – structural, 138, 144, 157, 160, 162, 163, 166, 167, 198, 240, 246, 265, 267-285, 291, 320-322
 - regime, 107
 - risk analysis, 156, 181, 184 212, 219, 226, 244
 - risk management, 46, 168 194-196, 198, 199, 294
 - routing (propagation), 57, 60, 63, 79, 119, 123-126, 128, 129, 131, 132, 145, 147, 171
 - simulation, 80, 151, 167
 - spring floods, 111-115, 117, 231
 - storage reservoirs, 233, 234, 246
 - superimposition of floods, 72, 171, 214
 - transboundary floods, 1-12, 20, 23-32, 33-44, 46-47, 57-68, 133-134, 171-182, 193-218, 231, 239, 248, 259-264, 277-285, 287-304, 305-314, 315-333

flood(s) - vulnerability, 194
- warning, 70, 80, 196, 197, 200, 201, 288, 324
fuzzy logic, 224-226, 229
Geographical Information Systems (GIS), 2, 9, 11, 45, 46, 48, 55, 77, 101, 208, 209, 211-213, 216, 301, 302, 305-314
hydrological and meteorological ensembles, 57, 58, 62, 64, 67,
information management, 204
Körös Rivers, 24-26, 45-47, 52, 205-206, 213, 265, 268-275, 289
lag factor, 121, 132, 200
land cover, 45, 48, 99, 157, 208, 211, 212
land use - 48, 54, 102, 105, 209, 212, 295, 297, 307, 324
- change of, 99-101, 103, 105-109, 160, 166, 246
Markov chain, 171-173, 180
models/modelling, 197
    coupling of, 73, 75
    good application practice, 183-192
    HEC-RAS, 80, 214, 261
    Hydraulic, 133, 134, 138, 141, 161, 184, 301
    Hydrodynamic, 156, 161, 167
    hydrological (including forecasting), 60, 61, 67, 69, 70, 73, 83, 208, 214, 301
    mathematical/numerical, 155, 164, 185
    meteorological forecasting, 59, 61, 67-69, 71, 75, 76, 177
    quality assurance, 183, 185
    rainfall-runoff, 61, 69, 72, 73, 79, 83, 84, 100, 101, 184, 186
    verification, 143

MODIS, 1-11, 45, 48, 50-53, 209, 210, 293, 299, 300
Muskingum method (flood routing), 121, 122, 214
NATO, 25, 45, 46, 55, 205, 206, 208, 287-303, 305-306
Polders, 211, 319, 321
precipitation forecast, 60
remote sensing, 1-11, 45-55, 287-305
river discharge - longitudinal discharge profile, 16
    - measurement, 20-22
    - rating curve, 18, 19
runoff - components, 103, 109
    - depth, 104-106, 111-115, 118, 119
    - formation, 92, 113, 207, 290
    - maximum, 91, 96, 111-113, 117-119, 215, 263
rural catchments, 104, 157
sensors, 8, 37, 39
shallow water equations, 143, 144, 145, 152
telemetry - centre, 28
    - hydrological system, 23, 25-31, 34-43, 82
    - retranslation station, 36, 39
uncertainty, 184, 203, 220, 222
unsteady flow, 133-142, 261
urban areas, 143, 148, 238, 240
Water Framework Directive, 184
water management, 185, 186
weather forecasts (numerical), 58-60, 70, 71, 79, 81, 82, 85-87

Printed in the United States
84653LV00001B/82/A